# ASTRONOMY FOR TEENAGERS

Prof. Robert Stewart

**Publish Drive and V.F. Walker**

Copyright © 2024 Prof. Robert Stewart

All rights reserved

The characters and events portrayed in this book are fictitious. Any similarity to real persons, living or dead, is coincidental and not intended by the author.

No part of this book may be reproduced, or stored in a retrieval system, or transmitted in any form or by any means, electronic, mechanical, photocopying, recording, or otherwise, without express written permission of the publisher.

ISBN: 9798879333855

Cover design by: Markee Books
Library of Congress Control Number: 2018675309
Printed in the United States of America

# CONTENTS

Title Page
Copyright
Prologue
Astronomy For Teenagers   1
Extraterrestrial Life   3
Dark Matter and Dark Energy   11
Dark Matter   13
Dark Energy   17
The Essence of Dark Matter and Dark Energy   21
Famous Astronomers and Physicists   34
Acharya Kanada   35

| | |
|---|---:|
| Albert Einstein | 40 |
| Hasan Ibn Al-Haytham | 44 |
| Francis Bacon | 48 |
| Nicolaus Copernicus | 53 |
| René Descartes | 57 |
| Galileo Galilei | 66 |
| Sir. Isaac Newton | 71 |
| Johannes Keppler | 77 |
| Max Planck | 81 |
| James Clerk Maxwell | 85 |
| Edwin Hubble (The Father of Modern Cosmology) | 89 |
| Wernher von Braun | 95 |
| Werner Heisenberg | 101 |
| Paul Dirac | 107 |
| Erwin Schrodinger | 112 |
| Neils Bohr | 118 |
| Arthur Eddington | 123 |
| George Lemaitre | 128 |
| Richard Feynman | 134 |
| Wolfgang Pauli | 140 |
| Carl Sagan | 145 |
| Arthur C. Clark | 149 |
| Stephen Hawking | 154 |
| Michio Kaku | 159 |

| | |
|---|---|
| Alan Guth | 166 |
| Historical Black Scientists | 172 |
| Benjamin Banneker | 173 |
| Neil deGrasse Tyson | 191 |
| Katherine Johnson | 197 |
| Hakeem M. Oluseyi | 203 |
| Dorothy Vaughn | 209 |
| Gladys West | 214 |
| ABOUT THE AUTHOR | 220 |
| BIBLIOGRAPHY | 224 |

# PROLOGUE

Astronomy is an immensely captivating field of study. The universe offers a plethora of fascinating knowledge, ranging from the celestial bodies inside our solar system to the most remote corners of space. You may find it intriguing to acquire knowledge on the various classifications of stars, the evolutionary process of galaxies, or the investigation of extraterrestrial entities through space probes. Additionally, there is the chronicle of astronomy and the remarkable breakthroughs achieved by astronomers over the course of history.

I wrote this book because during my studies of *Astronomy at U.C. Berkeley* (earning an Sc.D. in the field), most of what was taught involved the discoveries of historical Astronomers; *nothing about their private lives*. Consequently, you'll learn about the scientists and their backgrounds.

The scientists in this book are *only a few of numerous* who have made noteworthy contributions to the fields of astronomy, physics, cosmology, astrophysics, and so on. My hope is that after reading the personal stories of these scientists, you will gain a burning desire *to discover more* about the countless other scientists and their contributions to the aforementioned fields. Perhaps, *you* will be the *next great contributor* to these fields, or at least, you'll look up to the night sky and say what I said as a toddler: *"I wonder what else is up there that I cannot see?"*

# ASTRONOMY

# FOR TEENAGERS

In a little village situated amidst the celestial bodies, resided an inquisitive young girl by the name of Janet. Each evening, she clandestinely exited her bedroom through the window and silently made her way to the fields, where she would gaze upwards at the radiant constellations. While extending her hand towards the shimmering lights, she had an unfamiliar sensation in her fingertips and instantaneously embarked on a fantastical expedition around the cosmos, encountering celestial bodies such as planets, stars, and a playful comet named Comitus. Accompany Janet as she sets forth on an extraordinary journey to uncover the mysteries of the universe and find her own position within the celestial realm.

# EXTRATERRESTRI

# AL LIFE

The inquiry into the existence of life on other celestial bodies is a highly significant and captivating enigma within the field of astronomy. Scientists have extensively investigated this potentiality for several decades, conducting thorough investigations to uncover substantiation of alien life within our solar system and beyond.

The presence of liquid water is a crucial determinant in assessing the possibility of life on distant worlds. Scientists search for celestial bodies, such as planets and moons, that potentially possess liquid water seas or underground water, as water is crucial for sustaining life as we understand it. Jupiter's moon Europa and Saturn's moon Enceladus are hypothesized to possess subglacial oceans beneath their ice exteriors, rendering them captivating subjects in the quest for alien life.

Another area of investigation is the examination of exoplanets, which are celestial bodies that revolve around stars beyond our solar system. Astronomers employ terrestrial telescopes and orbital observatories to detect exoplanets and investigate their atmospheres. Scientists want to detect signs of life by examining the chemical makeup of exoplanet atmospheres. Scientists are interested in investigating the conditions that have the potential to sustain life on other planets, in addition to hunting for indications of life. This encompasses comprehending the habitable zone encircling stars, wherein the circumstances are conducive for the presence of liquid water on the surface of a planet. The detection of exoplanets within the habitable zone has generated enthusiasm and revived curiosity in the quest for alien life. Although there is presently no conclusive evidence of the existence of life beyond Earth, scientists persist in their pursuit due to technological advancements and an expanding comprehension of the universe. The prospect of uncovering extraterrestrial life is an intriguing and inspirational domain in the field of astronomy, catching the imagination of both scientists and the general public.

Researchers have identified multiple celestial bodies, including planets and moons, that possess favorable conditions and attributes that make them viable candidates for supporting life. Notable objectives in the quest for alien life encompass:

1. <u>Mars</u>:

Has consistently captivated attention in the quest

to uncover evidence of the previous or current existence of life. Empirical data indicates that there is a possibility that liquid water existed on the outside of Mars at some point in time, and recent findings have revealed the existence of water ice beneath the surface. Future missions, including the Mars Perseverance rover and the Mars Sample Return mission, will persist in their pursuit of evidence of past or current microbial life on Mars, alongside previous missions like the Mars rovers.

2. <u>Europa, a moon of Jupiter</u>:

is postulated to possess a subterranean ocean located beneath its icy outer layer. Europa's possession of liquid water and the existence of hydrothermal vents make it an intriguing subject for investigating habitability and the potential discovery of life in its underground ocean.

3. <u>Enceladus, one of Saturn's moons</u>:

Is believed to possess a subterranean ocean, which has been observed through the eruption of geysers from the moon's southern pole. The presence of water vapor, organic molecules, and other chemicals in these geysers has aroused the curiosity of scientists in their quest to discover extraterrestrial life.

4. <u>Proxima Centauri b</u>:

Is situated within the habitable zone of Proxima Centauri, which is the nearest star to the Sun that we are aware of. Although there is still a lot of information to gather about this exoplanet, its close distance to Earth and its capacity to potentially sustain liquid water make it a captivating subject for future scientific investigations.

5. <u>The TRAPPIST-1 system</u>:

Situated 40 light-years from Earth, comprises seven exoplanets that are similar in size to our planet. Notably, three of these exoplanets are positioned within the habitable zone of their parent star. The identification of this system has sparked enthusiasm for investigating these exoplanets in order to ascertain their suitability for sustaining life and the existence of conditions conducive to supporting life.

These examples represent only a small selection of the numerous captivating objectives in the quest for alien life. Although there is considerable conjecture and enthusiasm around these celestial objects, additional research and scientific inquiry are necessary to ascertain the genuine possibility of extraterrestrial life. The continuous progress in technology and upcoming missions, along with the persistent study of exoplanets, will have pivotal significance in our pursuit to discover possible habitats for life in the universe.

# DARK MATTER AND DARK ENERGY

These two riddles are highly intriguing in the field of current astrophysics, as they make up a substantial part of the universe's mass-energy composition and have a profound impact on cosmic structures and dynamics.

# DARK MATTER

Dark matter is a mysterious type of stuff that lacks the ability to emit, absorb, or reflect light, rendering it opaque to conventional telescopes and impervious to detection through electromagnetic radiation. Dark matter's existence is deduced from its gravitational impact on the movement of galaxies, galaxy clusters, and the overall structure of the universe, even if its true nature remains mysterious. Below are few essential aspects about dark matter:

1. The primary manifestation:

Of dark matter is its gravitational influence. Due to its substantial bulk, dark matter exerts gravitational forces that unite galaxies and influence the movements of stars within galaxies, as well as the dynamics of galaxy clusters. Astronomers can indirectly map the distribution of dark matter in the universe by using gravitational lensing, which is the phenomenon of light being bent by the gravitational field of dark matter.

2. Composition:

Although the exact characteristics of dark matter are still uncertain, several theoretical frameworks suggest that dark matter could be composed of unconventional subatomic particles that exhibit weak interactions with ordinary matter and electromagnetic radiation. Experiments undertaken in underground laboratories and particle accelerators are actively searching for hypothetical particles, including Weakly Interacting Massive Particles (WIMPs) and axions.

3. Cosmological Significance:

Dark matter is crucial in the process of shaping and developing cosmic structures. The gravitational force it exerts facilitates the formation of expansive structures, including galaxy clusters and cosmic web filaments, which serve as the framework for the observed arrangement of galaxies in the universe.

# DARK ENERGY

<u>Dark energy</u> is an enigmatic and repulsive force believed to be responsible for the ongoing acceleration of the universe's expansion. Dark energy, in contrast to gravity, opposes the force of gravitational attraction on a cosmic scale, resulting in the accelerated expansion of the universe. Below are few crucial aspects about dark energy:

1. <u>Accelerated Expansion</u>:

Observations made in the late 1990s of distant supernovae indicated that the expansion of the universe is accelerating, contrary to expectations based only on the influence of gravity. The unanticipated increase in speed prompted the development of the notion of dark energy, which is hypothesized to permeate the cosmos and drive the expansion of the universe.

2. <u>The nature of dark energy</u>:

Whether it be in the form of a cosmological constant or quintessence, continues to be a major enigma in the field of astrophysics. The cosmological constant, which is a term in Einstein's equations of general relativity, offers one possible explanation for dark energy. It indicates a constant energy density of space itself. Alternatively, dark energy may originate from a dynamic or changing energy field, referred to as quintessence, which demonstrates qualities that fluctuate throughout time and space.

3. <u>Impact on the Universe</u>:

The existence of dark energy affects the destiny and spatial structure of the universe. If the phenomenon of dark energy persists in fueling the ongoing accelerated expansion of the universe, it has the potential to result in a catastrophic event known as the "Big Rip." In this scenario, the universe's expansion would accelerate to such an extreme extent that it would eventually cause the disintegration of cosmic structures, including galaxies and even atoms, in the far future.

Although dark matter and dark energy are separate phenomena, they both play a role in the intricate process of cosmic evolution and have significant consequences for our comprehension of the universe's history, current state, and future. The investigation of dark matter and dark energy is a prominent area of study in astrophysics, particle physics, and cosmology, aiming to provide fundamental knowledge about the nature of the universe.

# THE ESSENCE OF DARK MATTER AND DARK ENERGY

Scientific exploration and theoretical conjecture persist in the pursuit of understanding the mysteries of dark matter and dark energy. Although conclusive solutions are yet unknown, scholars have proposed multiple new theories and lines of inquiry to better understand the characteristics of these enigmatic cosmic components. Below are the latest advancements and theoretical principles in the continuous pursuit to comprehend dark matter and dark energy:

1. <u>Theories of Modified Gravity</u>:

As a response to the enigmas presented by dark matter and dark energy, several scientists have investigated alternate theories of gravity that might potentially eliminate the necessity for these mysterious phenomena. Modified gravity theories, such as Modified Newtonian Dynamics (MOND) and Modified Gravity (MOG), suggest alterations to the laws of gravity at cosmic scales in order to explain observable events without the need for dark matter. Similarly, theories such as f(R) gravity and scalar-tensor theories propose alternative descriptions of gravity that differ from ordinary general relativity. These theories have the ability to account for the accelerated expansion of the universe caused by dark energy, without requiring any further explanations.

2. <u>Self-interacting dark matter</u>:

The behavior of dark matter at tiny scales, specifically inside galaxies and galactic clusters, has generated interest in models of self-interacting dark matter (SIDM). These theoretical frameworks suggest that dark matter particles have the ability to interact with each other through forces other than gravity. This interaction could result in observable effects, such as the creation of dark matter cores in small galaxies and changes to the way dark matter is distributed in clusters of galaxies.

3. <u>Alternative forms of axions and WIMPs</u>:

Due to ambiguous results from experimental searches for dark matter particles, specifically weakly interacting massive particles (WIMPs) and axions, researchers have investigated modified versions of these hypothetical particles with different characteristics and levels of interaction. An instance of this is the "fuzzy dark matter" hypothesis, which suggests that dark matter could be composed of extremely light bosonic particles with De Broglie wavelengths at the scale of galaxies. This could potentially impact the composition of galactic cores and the creation of dwarf galaxies.

4. <u>Emergent or Holographic Dark Energy</u>:

Several theoretical frameworks have attempted to reconcile the concept of dark energy with the fundamental laws of quantum field theory and gravity. These frameworks propose that dark energy could potentially arise from underlying quantum effects or holographic principles. These ideas suggest that the energy linked to dark energy does not originate from space itself, but rather arises from quantum entanglement or holographic information stored on cosmic horizons. This results in a perceived repulsive force that causes the universe to expand at an accelerated rate.

5. <u>Diverse Strategies</u>:

Researchers have employed diverse strategies to comprehend the enigmatic nature of dark matter and dark energy, acknowledging their intricate and intertwined relationship with bigger phenomena in astrophysics and particle physics. The goal of cross-collaboration between astrophysical observations, particle physics experiments, cosmological simulations, and theoretical modeling is to combine different types of data and limitations in order to understand the characteristics of dark matter and dark energy. Although the aforementioned instances serve as only illustrations of contemporary theoretical conceptions and advancements pertaining to dark matter and dark energy, it is crucial to emphasize the dynamic and continuous evolution of the scientific investigation into these enigmas. The investigation into the genuine characteristics of dark matter and dark energy continues to be a crucial area of study in astrophysics and cosmology. This pursuit has the capacity to fundamentally transform our comprehension of the universe and the underlying principles that dictate its development. Continuing and upcoming observations, experiments, and theoretical inquiries offer potential for elucidating these fundamental cosmic mysteries.

Modified gravity theories are a fascinating set of theoretical frameworks that question the conventional understanding of general relativity and aim to offer alternative explanations for how gravity operates on large cosmic sizes. These ideas are formulated to explain astrophysical phenomena, such as the rapid expansion of the universe and the behavior of galaxies and galactic clusters, without relying on the presence of dark matter or dark energy. Below is an elaborate exposition on modified gravity theories and its fundamental concepts:

## 1. Modified Newtonian Dynamics (MOND):

Is a theoretical framework that proposes a modification to Newton's laws of motion. Proposed in the 1980s by Mordehai Milgrom, MOND (Modified Newtonian Dynamics) proposes alterations to the principles of gravity under situations of low acceleration. This theory deviates from the conventional Newtonian dynamics and general relativity. MOND posits that in low-acceleration environments, such as within galaxies, gravitational forces exhibit deviations from the predictions of classical physics. MOND introduces a critical acceleration scale, below which gravity shows nonlinear behavior. This successfully alters the force law to explain observed galactic rotation curves without the need for huge amounts of dark matter.

## 2. Modified Gravity (MOG):

Refers to a theoretical framework that proposes alternative equations of gravity to those described by Einstein's general theory of relativity. Modified Gravity, also known as MOG, is a different approach that suggests making changes to the structure of space-time and the way gravity works. This is done by adding other fields or degrees of freedom, in addition to the metric tensor and its related curvature, which are the main components in general relativity. Within the framework of Modified Gravity (MOG), the gravitational force law is altered when dealing with significant distances. This modification provides an alternate interpretation for gravitational phenomena observed on cosmological scales, without the need to invoke the existence of dark matter or dark energy.

3. <u>Gravity theories involving f(R) and scalar-tensor</u>

<u>theories</u>:

These theoretical frameworks suggest alterations to the theory of general relativity by incorporating extra scalar fields that interact with the gravitational field. In f(R) gravity, the conventional Einstein-Hilbert action in general relativity is replaced by a more comprehensive gravitational action that incorporates nonlinear functions of the Ricci scalar R. Scalar-tensor theories propose the presence of scalar fields that interact with the metric tensor and alter the gravitational dynamics on a vast scale. Modified gravity theories seek to explain the accelerated expansion of the universe and the behavior of cosmic structures. They achieve this by modifying the equations that describe the gravitational field, allowing for additional variables beyond the usual metric tensor.

## 4. Chameleon and Symmetron Mechanisms:

Some modified gravity theories feature mechanisms that allow for the effective suppression or augmentation of new gravitational interactions in high-density situations, such as within the proximity of huge masses. For example, the chameleon mechanism adds a scalar field that is highly sensitive to the local matter density, resulting to a screening effect that suppresses the additional gravitational force associated with the scalar field in places of high matter density.

Similarly, the symmetron mechanism utilizes scalar fields that exhibit variable behaviors based on the ambient matter density, effectively modifying the strength of the related scalar-tensor gravitational interactions in different situations. Modified gravity theories have received interest due to their potential to address cosmic phenomena while evading the necessity for exotic forms of matter and energy, such as dark matter and dark energy. These theoretical frameworks are actively studied through astrophysical observations, cosmological simulations, and experimental tests, embracing a wide spectrum of phenomena that collectively probe the validity and predictions of modified gravity theories.

It is crucial to emphasize that while modified gravity theories offer intriguing alternatives to standard general relativity, they are subject to rigorous empirical inspection and verification. Experimental tests, including as precision measurements of gravitational events, cosmological observations, and laboratory experiments, play a significant role in evaluating the validity of these ideas and separating them from existing frameworks. Moreover, modified gravity theories constitute a fertile ground for multidisciplinary research, crossing with astrophysics, cosmology, particle physics, and fundamental principles of gravity and quantum mechanics. Their promise to illuminate the underlying nature of gravitational interactions on cosmic scales makes them an intriguing field of theoretical and observational investigation within the broader context of our understanding of the universe's fundamental features and evolution.

# FAMOUS ASTRONOMERS AND PHYSICISTS

# ACHARYA

# KANADA

Acharya Kanada, also known as Kashyapa, was an ancient Indian philosopher who made significant contributions to the realm of philosophy and science. He is well recognized for his work in the subject of atomic theory, which set the framework for the creation of modern physics. This writing tries to explore the contributions of Acharya Kanada, studying his atomic theory and its consequences for the knowledge of the physical universe.

Acharya Kanada's atomic hypothesis, known as <u>Vaisheshika</u>, is one of the first recorded theories on the nature of matter. According to this view, all matter is formed of microscopic, indivisible components called atoms. These atoms are eternal and indestructible, and they mix in different ways to form various substances. Kanada argued that atoms had distinct properties, such as taste, smell, color, and touch, which govern their behavior and interactions.

One of the essential aspects of Kanada's atomic theory is the concept of "paramanu," which refers to the smallest indivisible particle. Kanada hypothesized that these paramanus merge to produce larger particles, ultimately contributing to the genesis of matter. This concept is akin to the contemporary conception of atoms as the building units of matter. Kanada's theory also indicated that the properties of substances come from the arrangement and combination of atoms, providing a fundamental explanation of the physical world.

Acharya Kanada's atomic theory has various consequences for the understanding of the physical world. Firstly, his idea questioned the conventional belief in ancient India that matter was continuous and infinitely divisible. Kanada's view of atoms as separate particles gave a new perspective on the nature of matter, which corresponded with the present understanding of atomic theory.

Furthermore, Kanada's atomic theory established the framework for the creation of contemporary physics. His insights on the characteristics and interactions of atoms provided a framework for further investigation and experimentation in the topic. Although Kanada's atomic theory was not as comprehensive as the later theories proposed by scientists like John Dalton and Niels Bohr, it played a key part in defining the knowledge of matter and its behavior.

Acharya Kanada's contributions to the field of philosophy and science, particularly his atomic theory, have had a lasting impact on our knowledge of the physical universe. His theories challenged established views and formed the groundwork for the development of modern physics.

# ALBERT EINSTEIN

Albert Einstein, one of the most prominent and beloved men in the history of science, made profound and enduring contributions to theoretical physics, altering our knowledge of space, time, and the nature of the cosmos itself. His breakthrough work, including the formulation of the theory of general relativity and the theoretical foundation of the photoelectric effect, radically transformed the landscape of modern physics and established the platform for many advancements in cosmology, astrophysics, and quantum theory.

Einstein's <u>theory of general relativity</u> (published in 1915), presented a completely new notion of gravity. Where Newton's laws portrayed gravity as a force that acted at a distance, Einstein's general relativity reinvented gravity as the curvature of space-time induced by the presence of matter and energy. This monumental theory provides a unified picture of gravity and space-time, bringing fresh insights into the behavior of huge objects, the bending of light in gravitational fields, and the dynamics of the universe at large scales.

Einstein's discovery on the photoelectric effect, for which he was awarded the Nobel Prize in Physics in 1921, played a vital role in the development of quantum theory. His theoretical explication of the quantization of light and the concept of photons helped create the framework for the burgeoning discipline of quantum mechanics, dramatically affecting subsequent study in particle physics and the fundamental nature of matter and radiation.

Beyond his scientific achievements, Einstein's intellectual and philosophical effect extended to his advocacy for pacifism, civil rights, and scientific cooperation, showing a fundamental dedication to the ethical and social dimensions of scientific investigation. Einstein's long legacy continues to inspire and guide generations of scientists, underlining the transformational power of visionary intellect, creative imagination, and a dogged search of truth in solving the secrets of the universe.

His profound discoveries and enduring contributions have left an indelible stamp on the fabric of scientific understanding, resonating across the boundaries of theoretical physics, cosmology and the broader human quest for knowledge and understanding.

# HASAN IBN AL-HAYTHAM

Alhazen, also known as Ibn Al-Haytham, (c. 965 – c. 1040 CE) discovered a formula for the sum of fourth powers. He utilized the results to conduct out what would now be called an integration, where the formulas for the sums of integral squares and fourth powers allowed him to compute the volume of a paraboloid. Thus, giving rise to the *earliest form of calculus*. He was a pioneering scientist and polymath who made outstanding contributions to several subjects, including optics, physics, and mathematics. He lived during the Islamic Golden Age, making great discoveries in the study of light, vision, and the fundamentals of optics.

One of Alhazen's most influential works is his thorough and revolutionary study of optics, <u>Kitab al-Manazir</u> (The Book of Optics). This foundational treatise, authored in *the 11th century*, substantially affected the development of optics and visual perception in both the Eastern and Western worlds. In this massive work, Alhazen exhaustively examined the behavior of light, the anatomy of human eyesight, and the physics of reflection and refraction, yielding important insights that lay the groundwork for future developments in optics.

Alhazen is recognized for his experimental approach and emphasis on actual observation, distinguishing his *scientific methodology* from prevailing speculative and theoretical frameworks. His thorough experimentation and empirical inquiries into light and vision predicted the later scientific procedures that would become crucial to modern experimental science. Hence, he is <u>the originator</u> of "<u>The Scientific Method</u>."

Furthermore, Alhazen's work on the

camera-obscura, a forerunner to contemporary cameras, and his studies of spherical and parabolic mirrors revealed his pioneering contributions to the understanding of optics and the principles of light manipulation. Alhazen's wide influence spread to other subjects, including mathematics and physics. His work on the principles of motion and the rules of reflection have left a lasting mark on the domains of classical mechanics and geometry.

Al-Haytham's work acted as a bridge between ancient and modern science, dramatically influencing later scholars and affecting the direction of scientific inquiry and discoveries unto this day.

# FRANCIS BACON

An English philosopher, statesman, scientist, and lawyer, who is largely recognized as one of the most influential figures in the formation of modern or *Western Scientific methodology*. As previously established, *Ibn Al-Haytham* is the *original father of the Scientific Method*. Francis only altered Al-Haytham's pioneering Scientific Method, nothing more.

Born in 1561, Bacon made enduring contributions to several subjects, with his impact most profoundly felt in the realms of philosophy, empirical investigation, and the advancement of the Scientific Method. Bacon's seminal work <u>Novum Organum</u> (published in 1620) stands as a landmark treatise that challenged prevailing modes of scientific investigation and laid the groundwork for a new approach to empirical inquiry. In this foundational work,

Bacon articulated his vision for a systematic methodology of scientific investigation based on inductive reasoning and rigorous empirical observation. He promoted the idea that scientific knowledge should be built on a solid foundation of *empirical evidence, verifiable experimentation, and the careful collection of data, free from preconceived notions or biases*. Bacon's influential advocacy for the systematic and methodical pursuit of knowledge through empirical observation and experimentation (first established by Ibn Al-Haytham) marked a significant departure from the speculative and deductive traditions of knowledge acquisition prevalent in his era and in the Western Hemisphere. His emphasis on the primacy of observation and the systematic testing of hypotheses underscored his commitment to the principles of empirical inquiry and the rigorous examination of natural phenomena. Another persistent part of Bacon's intellectual legacy resides in his explication of the concept of the "four idols" that inhibit clear and objective reasoning:

1. the idols of the tribe.
2. the idols of the cave.
3. the idols of the marketplace.
4. the idols of the theater.

These idols symbolize the numerous kinds of human mistake and bias that can distort thinking and block the pursuit of genuine truth. Bacon's perceptive analysis of these barriers illustrates his profound insight into the limitations of human cognition and the quest for unbiased scientific investigation. Bacon's far-reaching influence extended beyond his contributions to the philosophy of science. As a politician and lawyer, he left an indelible stamp on the realms of government, law, and public administration, defining his era's intellectual and political landscapes. The enduring resonance of Bacon's ideas and his profound impact on the evolution of scientific methodology continue to inspire contemporary scholars and scientists, underscoring the enduring relevance of his visionary insights into the principles of empirical investigation, rational inquiry, and the pursuit of knowledge.

# NICOLAUS COPERNICUS

This Renaissance mathematician and astronomer revolutionized our understanding of the cosmos with his heliocentric model of the solar system. His seminal work, <u>De revolutionibus orbiumcoelestium</u> (On the Revolutions of the Celestial Spheres), published in 1543, challenged the prevailing geocentric model, which posited Earth as the center of the universe, and proposed a sun-centered model that profoundly reshaped our perception of the cosmos.

Copernicus's heliocentric model proposed that the planets, including Earth, revolve around the Sun in circular orbits, marking a transformative departure from the ancient Ptolemaic system. By repositioning the Sun at the center of the solar system, Copernicus's model offered a more elegant and coherent framework for understanding the movements of celestial bodies, effectively laying the groundwork for the "Copernican Revolution" and ushering in a new era of scientific inquiry empirical observation, and mathematical understanding of the universe.

While Copernicus' heliocentric model faced initial resistance and scrutiny, particularly from religious and philosophical authorities, its enduring impact precipitated a paradigmatic shift in our comprehension of the cosmos. His revolutionary insights catalyzed advancements in observational astronomy, celestial mechanics, and the foundations of modern physics, establishing a legacy that continues to resonate across the frontiers of scientific discovery and cosmological understanding.

Copernicus's steadfast commitment to empirical observation, mathematical rigor, and visionary thinking serves as an enduring testament to the transformative power of scientific inquiry, inspiring subsequent generations of astronomers, physicists, and mathematicians in their pursuit of knowledge and understanding of the universe. His contributions stand as a testament to the indelible impact of perseverance, intellectual courage, and visionary insight in reshaping humanity's understanding of the natural world.

# RENÉ DESCARTES

A prominent French philosopher, mathematician, and scientist, who is celebrated for his profound contributions to Western philosophy, mathematics, and the development of modern thought. Born in 1596, Descartes made transformative strides in multiple disciplines, leaving an undeniable mark on the intellectual landscape, and significantly shaping the trajectory of modern philosophy and scientific inquiry.

Descartes' most renowned work, <u>Meditations on First Philosophy</u>, stands as a foundational treatise that profoundly influenced the course of philosophy. In this seminal work, Descartes presented his philosophical method, which centered on the rigorous application of doubt as a means of arriving at certain knowledge.

Through his famous assertion "Cogito, ergo sum" (I think, therefore I am), Descartes launched a radical philosophical inquiry that emphasized the individual's capacity for critical thought and the foundational certainty of one's own existence as a thinking being. Descartes' methodical skepticism and emphasis on systematic doubt laid the groundwork for a new approach to philosophical inquiry, one that sought to establish a coherent and indubitable foundation for knowledge. His methodological skepticism prompted a rigorous examination of the nature of reality, the veracity of knowledge claims, and the pursuit of clear and distinct truths that would form the basis of a secure system of knowledge.

In addition to his profound impact on philosophy, Descartes made significant contributions to mathematics, most notably through the development of "Cartesian Coordinates," a revolutionary mathematical framework that laid the foundation for Analytic Geometry. This innovation established a powerful link between algebra and geometry, transforming the study of mathematics and providing a vital tool for the advancement of scientific inquiry and technological innovation.

Descartes' multifarious contributions also extended to the sphere of Natural Science where his mechanistic vision of the universe and devotion to empirical investigation left an enduring impression on the formation of contemporary scientific philosophy. His mechanistic perspective of the natural world, typified by the application of mathematical principles to the study of physical phenomena, played a vital influence in creating the scientific paradigm of future generations. The ongoing importance of Descartes' ideas on the philosophy of mind, the nature of consciousness, the quest of knowledge, and the scientific method, continues to inspire scholars and researchers across numerous fields.

Descartes' Cartesian coordinate system has found wide-ranging contemporary applications across various fields, spanning mathematics, physics, engineering, computer science, and beyond. This foundational mathematical framework, which provides a systematic way to represent geometric shapes and equations algebraically, continues to underpin diverse areas of research and practical applications in the modern world. These include:

1. <u>Engineering and Architecture</u>:

In engineering and architecture, Cartesian coordinates serve as essential tools for geometric design, structural analysis, and spatial modeling.

They facilitate precise spatial representation, aiding in the planning, design, and construction of complex structures, infrastructure, and mechanical systems.

2. <u>Physics and Astronomy</u>:

Cartesian coordinates play a key function in expressing the

locations, motions, and interactions of celestial bodies and physical systems in space. They constitute the basis for celestial navigation, astrodynamics, and the mathematical modeling of physical processes, enabling study in astrophysics, orbital mechanics, and space exploration.

3. <u>Computer Graphics & Visualization</u>:

Cartesian coordinates are crucial to computer graphics, visual rendering, and geometric modeling. They form the basis for expressing digital images, three-dimensional objects, and graphical interfaces, enabling complex visualization techniques and virtual reality applications.

4. Geographic Information Systems (GIS):

In GIS applications, Cartesian coordinates provide a means of spatial referencing and mapping, facilitating the analysis and visualization of geographic data. They support geospatial analysis, urban planning, environmental modeling, and the representation of geographical features and spatial relationships.

5. Robotics and Control Systems:

Cartesian coordinates are integral to robotics, automation, and control systems, enabling precise spatial positioning and manipulation of robotic arms, articulated mechanisms, and automated manufacturing processes. They contribute to the mathematical frameworks used in robot kinematics, motion planning, and spatial control.

## 6. Mathematical Modeling and Simulation:

Cartesian coordinates are widely employed in mathematical modeling and simulation across scientific disciplines, including fluid dynamics, electromagnetism, and material science. They provide a standardized framework for representing and analyzing physical systems through mathematical equations and computational simulations.

## 7. Machine Learning and Data Analysis:

In machine learning and data analysis, Cartesian coordinates are instrumental in multivariate analysis, pattern recognition, and feature engineering. They facilitate the representation and analysis of complex datasets, supporting the development of algorithms and statistical methods for data-driven applications.

8. <u>Medical Imaging and Biomechanics</u>:

Cartesian coordinates are applied in medical imaging techniques such as MRI, CT scans, and 3D reconstruction of anatomical structures. They also enable biomechanical analysis, enabling the representation and quantitative assessment of movement, musculoskeletal dynamics, and physiological systems.

These examples emphasize the extensive importance of Cartesian coordinates in modern mathematics and science, where they continue to underlie a diverse variety of applications, from fundamental study to technological advancements.

# GALILEO GALILEI

Galileo Galilei, an Italian physicist, mathematician, and astronomer, is widely regarded as *one of the most influential figures* in the history of science. Born in 1564 in Pisa, Galileo's groundbreaking discoveries and innovative methods revolutionized our understanding of the natural world. This writing will analyze Galileo's contributions to science, focusing on his experimental approach, his discoveries in astronomy, and his conflict with the Catholic Church.

Galileo Galilei was a pioneer in the use of experimental methods in scientific inquiry. He believed that knowledge should be based on empirical evidence rather than relying solely on philosophical or theological arguments. Galileo's experiments with inclined planes and falling bodies challenged the prevailing Aristotelian view of motion. By meticulously measuring the time it took for objects to fall, he demonstrated that their speed increased uniformly, contradicting Aristotle's belief that heavier objects fell faster. Galileo's emphasis on observation and measurement laid the foundation for the scientific method (derived from Hasan Ibn-Al-Haytham's pioneering method) which remains the cornerstone of modern scientific inquiry.

Galileo's most significant contributions were in the field of astronomy. Using a telescope, which he improved upon and refined, Galileo made groundbreaking observations that challenged the geocentric model of the universe. He discovered the four largest moons of Jupiter, now known as the Galilean moons, which provided evidence that celestial bodies could orbit something other than the Earth. This observation directly contradicted the prevailing belief that all celestial bodies revolved around the Earth. Galileo's observations of the phases of Venus also supported the heliocentric model proposed by Nicolaus Copernicus, further undermining the geocentric view.

Galileo's revolutionary ideas and discoveries brought him into direct conflict with the Catholic Church, which held a geocentric view of the universe. In 1616, the Church issued a decree condemning Copernicanism, which Galileo had openly supported. Despite this, Galileo continued to defend the heliocentric model in his writings, leading to his eventual trial by the Inquisition in 1633. Galileo was found guilty of heresy and was placed under house arrest for the remainder of his life. This conflict between science and religion highlighted the tension between established beliefs and the pursuit of knowledge, a theme that continues to resonate in contemporary society.

Galileo Galilei's contributions to science were truly revolutionary. His experimental approach and emphasis on empirical evidence laid the foundation for modern scientific inquiry. His discoveries in astronomy challenged long-held beliefs and paved the way for the acceptance of the heliocentric model of the universe. Despite his conflict with the Catholic Church, Galileo's legacy as a scientific pioneer remains intact. His work continues to inspire scientists and scholars to question established beliefs and push the boundaries of knowledge. Galileo Galilei will forever be remembered as a visionary who changed the course of scientific history.

# SIR. ISAAC NEWTON

Sir Isaac Newton, born on January 4, 1643, in Woolsthorpe, Lincolnshire, England, is widely regarded as one of the most influential scientists in history. His groundbreaking contributions to the fields of physics, mathematics, and astronomy revolutionized our understanding of the natural world. Newton's analytical mind and rigorous approach to scientific inquiry laid the foundation for modern physics and established him as a towering figure in the scientific community.

Newton's early life was marked by hardship and adversity. His father died just three months before his birth, leaving his mother to raise him alone. Despite these challenges, Newton displayed exceptional intellectual abilities from a young age. He attended the University of Cambridge, where he studied mathematics and physics, and it was during this time that he began to develop his groundbreaking theories.

One of Newton's most significant contributions was his formulation of the laws of motion. In his seminal work, <u>Mathematical Principles of Natural Philosophy</u>, published in 1687, Newton outlined three fundamental laws that govern the motion of objects. These laws, known as Newton's laws of motion, provided a comprehensive framework for understanding the behavior of objects in motion. They laid the groundwork for classical mechanics and became the cornerstone of physics for centuries to come. *Newton's laws of motion* were not only revolutionary but also highly analytical in nature. He approached the study of motion with a mathematical mindset, using rigorous mathematical equations to describe the relationships between forces, mass, and acceleration. This analytical approach allowed him to make precise predictions about the behavior of objects, which could be tested and verified through experimentation.

In addition to his laws of motion, Newton also made significant contributions to the *field of optics*. He conducted experiments with light and developed the theory of color, demonstrating that white light is composed of a spectrum of colors. Newton's experiments with prisms and his subsequent publication, Opticks, published in 1704, laid the foundation for the field of optics (which was firmly established by Hasan Ibn Al-Haytham} and paved the way for future discoveries in the study of light.

Furthermore, Newton's work in mathematics was equally groundbreaking. He developed the *mathematical framework of calculus*

*(Newtonian Calculus)*, independently of German mathematician *Gottfried Wilhelm Leibniz*, and *Ibn Al-Haytham's calculus*. Calculus provided a powerful tool for solving complex mathematical problems and became an essential tool in physics and engineering. Newton's analytical approach to mathematics allowed him to solve problems that were previously unsolvable, opening up new avenues of scientific inquiry.

Newton's contributions to science were not limited to physics and mathematics. He also made significant advancements in the field of astronomy. His <u>law of universal gravitation</u>, published in the same work as his <u>laws of motion</u>, provided a mathematical description of the force that governs the motion of celestial bodies. This law allowed scientists to accurately predict the motion of planets and other celestial objects, revolutionizing our understanding of the universe.

In conclusion, Sir Isaac Newton's analytical mind and rigorous approach to scientific inquiry revolutionized our understanding of the natural world. His laws of motion, theory of optics, development of calculus, and law of universal gravitation laid the foundation for modern physics and established him as one of the greatest scientists in history. Newton's analytical approach to science, coupled with his exceptional intellect, allowed him to make groundbreaking discoveries that continue to shape our understanding of the universe. His legacy as the father of modern physics will forever be remembered and celebrated.

# JOHANNES

# KEPPLER

This pioneering astronomer and mathematician made remarkable contributions to our understanding of the cosmos during the 17th century. Kepler's groundbreaking work profoundly influenced the development of modern astronomy and laid the foundation for the scientific revolution that transformed our perception of the universe.

Kepler is celebrated for his formulation of <u>the three laws of planetary motion</u>, known as Kepler's laws, which revolutionized our understanding of the dynamics governing the movements of celestial bodies. His first law, also called the law of ellipses, established that planets move in elliptical orbits with the Sun, departing from the prevailing belief in circular orbits. The second law, the law of equal areas, demonstrated that a line segment joining a planet and the

Sun sweeps out equal areas in equal intervals of time, providing critical insights into the speed of planetary motion at different points in their orbits. Kepler's third law, the harmonic law, established a quantitative relationship between a planet's orbital period and its average distance from the Sun, offering a fundamental connection between the dynamics of celestial motion and the properties of the planetary system. Kepler's tireless pursuit of empirical observations, coupled with his profound mathematical expertise, empowered him to uncover these fundamental principles that govern the movements of planets around the Sun, engendering a transformative shift in our comprehension of the solar system.

His principles have stood the test of time, continuing to serve as vital tools for astronomers and astrophysicists in understanding the behavior of planetary systems, including those beyond our own solar system.

Beyond his contributions to planetary motion, Kepler's work extended to diverse fields of study, including optics and the development of *the Keplerian telescope*. His impact reverberates throughout the annals of scientific inquiry, embodying the power of empirical observation, mathematical rigor, and visionary insight in reshaping our understanding of the natural world. Kepler's legacy survives as an emblem of scientific enthusiasm and inventiveness, inspiring generations of astronomers, physicists, and mathematicians in their search of knowledge and comprehension of the cosmos.

# MAX PLANCK

A distinguished physicist who is recognized for *his revolutionary contributions to quantum theory*. He stands as a towering figure in the annals of scientific inquiry. Born in 1858 in Kiel, Germany, Planck's pioneering work altered the course of theoretical physics, fundamentally transforming our understanding of the fundamental properties of matter and radiation.

One of Planck's most enduring achievements is his formulation of the concept of *energy quantization*, which served as the cornerstone of quantum theory. In 1900, Planck introduced the revolutionary idea that energy, in the form of electromagnetic radiation, is emitted or absorbed in discrete, indivisible units known as "*quanta*." This groundbreaking postulate encapsulated in *Planck's equation $E = hf$*, where E represents energy, h signifies Planck's constant, and f denotes the frequency of the radiation, heralded a paradigm shift in our comprehension of the nature of light and energy.

Planck's pivotal insight reconciled experimental observations with theoretical predictions, effectively resolving long-standing discrepancies that had confounded physicists by introducing the radical notion of *quantized energy*, Planck laid the foundation for the development of *quantum mechanics*, igniting a revolution in physics that would reshape our understanding of the microscopic realm and underpin the evolution of modern physics.

Planck's epochal contributions extended beyond the realm of quantum theory, encompassing diverse branches of theoretical and applied physics. His incisive intellect and profound mathematical acumen made indelible marks in fields such as thermodynamics, statistical mechanics, and radiation physics, cementing his status as one of the most influential physicists of the 20th century.

# JAMES CLERK MAXWELL

This pioneering Scottish Physicist made revolutionary advances to our understanding of *electromagnetism* and the unification of the fundamental forces of existence. Born in 1831, Maxwell's profound ideas and mathematical acumen revolutionized theoretical physics, setting the framework for the creation of modern *electromagnetic theory and quantum mechanics*.

One of Maxwell's most enduring achievements is the formulation of a concise set of equations (known as <u>Maxwell's equations</u>) which elegantly describe the behavior of electric and magnetic fields and their dynamic interplay. These foundational equations united and codified the principles of electricity, magnetism, and light, providing a unifying framework that comprehensively explained a wide range of electromagnetic phenomena. Maxwell's equations have since been crucial in numerous domains, from telecommunications and electrical engineering to theoretical physics and cosmology, illustrating the ongoing effect of his seminal contributions.

Central to Maxwell's theoretical framework was his groundbreaking proposal that light is an electromagnetic wave, elaborated in his celebrated treatise <u>A Dynamical Theory of the Electromagnetic Field</u>. This pivotal work not only revolutionized our understanding of light, but also paved the way for the development of *quantum theory* and the modern conception of the electromagnetic spectrum, encompassing radio waves, microwaves, infrared radiation, visible light, ultraviolet radiation, X-rays, and gamma rays.

Maxwell's deep insights into the linked

nature of electricity magnetism, and light significantly revolutionized our view of the natural world, heralding a paradigm shift in theoretical physics and igniting a renaissance in experimental inquiry and technological innovation. James Clerk Maxwell's important contributions to theoretical physics, electromagnetism and the unification of fundamental forces emphasize his place as a leading figure in the pantheon of scientific geniuses.

# EDWIN HUBBLE (THE FATHER OF MODERN COSMOLOGY)

Edwin Hubble, an American astronomer, is widely regarded as one of the most influential figures in the field of cosmology. His groundbreaking discoveries and contributions to our understanding of the universe have shaped the way we perceive and study the cosmos. We will analyze the life, work, and impact of Edwin Hubble, highlighting his significant contributions to the field of astronomy.

Edwin Powell Hubble was born on November 20, 1889, in Marshfield, Missouri. He showed an early interest in science and astronomy, which led him to pursue a career in the field. Hubble completed his undergraduate studies at the University of Chicago, where he excelled in mathematics and astronomy. He then went on to earn his Ph.D. in astronomy from the University of Chicago in 1917.

Hubble's most notable contribution to astronomy was *his discovery of the expanding universe.* In the 1920s, using the 100-inch Hooker telescope at the Mount Wilson Observatory in California, Hubble observed distant galaxies and noticed a correlation between their redshifts and their distances from Earth. This observation led him to formulate *Hubble's Law*, which states that the velocity at which a galaxy is moving away from us is directly proportional to its distance from us. This groundbreaking discovery provided evidence for the expanding universe and supported the Big Bang theory.

Furthermore, Hubble classified galaxies into different types based on their shapes, such as spiral, elliptical, and irregular. His classification system, known as the Hubble sequence, is still widely used today. Hubble also played a crucial role in determining the size and age of the universe. By measuring the distances to various galaxies, he was able to estimate the age of the universe to be around 13 billion years.

Edwin Hubble's contributions revolutionized the field of cosmology and laid the foundation for our current understanding of the universe. His discovery of the expanding universe challenged the prevailing belief in a static and unchanging cosmos. Hubble's work provided the first observational evidence for the Big Bang theory (which is still just a hypothesis, not proven fact), which is now widely accepted as the most plausible explanation for the origin of the universe. Hubble's classification system for galaxies has been instrumental in studying the formation and evolution of galaxies. It has allowed astronomers to categorize and study galaxies based on their shapes, leading to a deeper understanding of their properties and characteristics.

Furthermore, Hubble's determination of the age of the universe has had a profound impact on our understanding of its history. By estimating the age to be around 13 billion years, Hubble provided a timeline for the evolution of the cosmos, shedding light on the formation of stars, galaxies, and other celestial objects. Edwin Hubble's contributions to the field of cosmology have been invaluable. His discovery of the expanding universe, classification system for galaxies, and determination of the age of the universe (which has been *debunked* by the Webb telescope which now has the universe at *28 billion years old*) have shaped our understanding of the cosmos. Hubble's work continues to inspire and guide astronomers in their quest to unravel the mysteries of the universe. His legacy as the father of modern cosmology will forever be remembered and celebrated in the scientific community.

# WERNHER

# VON BRAUN

Wernher von Braun, born on March 23, 1912, in Wirsitz, Germany, was a renowned aerospace engineer and space architect. He played a pivotal role in the development of rocket technology and was instrumental in the United States' space program during the mid-20th century. This writing aims to analyze the life and contributions of Wernher von Braun, highlighting his achievements, controversies, and lasting impact on space exploration.

Wernher von Braun's interest in space exploration began at an early age. Fascinated by the works of Jules Verne and Hermann Oberth, he started experimenting with rockets in his teenage years. Von Braun pursued his passion by studying mechanical engineering at the Berlin Institute of Technology, where he earned his doctorate in physics in 1934. During his studies, he became a member of the German Society for Space Travel, which further fueled his enthusiasm for rocketry.

Von Braun's most significant contribution to rocketry was his work on the V-2 rocket during World War II. He led a team of engineers at the Peenemünde Army Research Center, where they developed the V-2, the world's first long-range guided ballistic missile. Although the V-2 was primarily used as a weapon during the war, von Braun envisioned its potential for peaceful space exploration.

While von Braun's contributions to rocketry were groundbreaking, his involvement with the Nazi regime during World War II remains a controversial aspect of his life. As a member of the Nazi Party and the SS, von Braun's work on the V-2 rocket was directly linked to Hitler's war efforts. After the war, von Braun and his team were captured by the Allies and brought to the United States under [Operation Paperclip](), where they continued their research under American supervision. Von Braun's association with the Nazi regime raises ethical questions regarding his moral responsibility. Some argue that he was merely a scientist forced to work for the Nazis, while others criticize him for willingly participating in the war effort. This controversy surrounding von Braun's past remains a topic of debate among historians and ethicists.

Despite the controversies surrounding his early career, von Braun's contributions to space exploration cannot be overlooked. After his arrival in the United States, he played a crucial role in the development of the American space program. He became the director of NASA's Marshall Space Flight Center and was instrumental in the design and development of the Saturn V rocket, which propelled the Apollo missions to the Moon.

Von Braun's vision and technical expertise were instrumental in shaping the future of space exploration. His advocacy for peaceful space exploration and his efforts to popularize the idea of space travel inspired generations of scientists and engineers. His work laid the foundation for subsequent advancements in rocket technology, making space exploration a reality. Wernher von Braun's contributions to rocketry and space exploration were undeniably significant. While his association with the Nazi regime remains a controversial aspect of his life, his technical expertise and visionary ideas propelled the United States to the forefront of space exploration. Von Braun's legacy continues to inspire scientists and engineers, reminding us of the importance of pushing the boundaries of human knowledge and exploration.

# WERNER HEISENBERG

Werner Heisenberg (born December 5, 1901 – February 1, 1976) was a German physicist who is widely known for his contributions to *quantum mechanics*, particularly his formulation of <u>the uncertainty principle</u>. This principle revolutionized our understanding of the physical world, challenging classical notions of determinism and introducing a fundamental limit to our ability to measure certain properties of particles. This writing aims to analyze Heisenberg's uncertainty principle, its implications, and its significance in the field of physics.

Heisenberg's uncertainty principle states that it is impossible to simultaneously determine the precise position and momentum of a particle with absolute certainty. In other words, the more accurately we measure the position of a particle, the less accurately we can determine its momentum, and vice versa. This principle arises from *the wave-particle duality of quantum mechanics*, where particles exhibit both wave-like and particle-like properties.

To understand the uncertainty principle, we must delve into the mathematical formulation. Heisenberg derived his principle using *matrices and operators*, which are fundamental tools in quantum mechanics. By considering the commutation relations between position and momentum operators, he showed that their product is subject to a lower limit, known as *Planck's constant*. This implies that the more precisely we measure one variable, the less precisely we can measure the other. The uncertainty principle has profound implications for our understanding of the physical world.

Firstly, it challenges the classical notion of determinism, which assumes that the position and momentum of a particle can be precisely known at any given time. Heisenberg's principle introduces an inherent randomness into the behavior of particles, highlighting the *probabilistic nature of quantum mechanics*.

Furthermore, the uncertainty principle has practical consequences. It sets a fundamental limit to the precision of measurements, as there will always be an inherent uncertainty in any measurement involving position and momentum. This limitation has implications in various fields, such as atomic and molecular physics, where precise measurements are crucial for understanding the behavior of particles at the quantum level. The uncertainty principle also has philosophical implications. It raises questions about the nature of reality and the limits of human knowledge. Heisenberg himself pondered the philosophical implications of his principle, suggesting that *our ability to observe and measure the physical world is inherently limited by the very act of observation*. This notion challenges the idea of an *objective reality independent of the observer*.

The significance of Heisenberg's uncertainty principle extends beyond its immediate implications. It paved the way for the development of quantum mechanics as a distinct branch of physics, revolutionizing our understanding of the microscopic world. The uncertainty principle, along with other principles of quantum mechanics, forms the foundation of modern physics and has led to numerous technological advancements, such as the development of quantum computing.

Werner Heisenberg's uncertainty principle has had a profound impact on the field of physics. Its formulation challenged classical notions of determinism and introduced a fundamental limit to our ability to measure certain properties of particles. The uncertainty principle has practical, philosophical, and scientific implications, shaping our understanding of the physical world and paving the way for the development of quantum mechanics. Heisenberg's contributions continue to inspire and influence physicists to this day, highlighting the importance of his work in shaping our understanding of the universe.

# PAUL DIRAC

Paul Dirac, a British theoretical physicist, is widely regarded as *one of the most influential scientists of the 20th century*. His groundbreaking contributions to the field of quantum mechanics revolutionized our understanding of the fundamental laws governing the universe. This writing aims to analyze Dirac's life, his scientific achievements, and his lasting impact on the field of physics.

Paul Adrien Maurice Dirac was born on August 8, 1902, in Bristol, England. From an early age, Dirac displayed exceptional mathematical abilities, which eventually led him to pursue a degree in engineering at the University of Bristol. However, his true passion lay in theoretical physics, and he soon switched his focus to this field. Dirac's most significant contribution to physics came in 1928 when he formulated the Dirac equation, which combined quantum mechanics and special relativity. This equation described the behavior of relativistic electrons and predicted the existence of antimatter. It was a groundbreaking achievement that laid the foundation for the development of *quantum field theory*.

In 1933, Dirac received the Nobel Prize in Physics with Erwin Schrödinger for their discovery of new productive forms of atomic theory. This acknowledgment confirmed Dirac's standing as a key figure in the scientific world. Dirac's work extended beyond the theoretical realm. He made significant contributions to the field of *quantum electrodynamics*, which describes the interaction between light and matter. His formulation of the quantum electrodynamics theory, known as the "hole theory," introduced the concept of antiparticles and provided a mathematical framework for understanding the behavior of particles and antiparticles.

Dirac's contributions to physics had a profound impact on the field, shaping the way we understand the fundamental laws of nature. His prediction of antimatter was confirmed experimentally, leading to the *discovery of the positron*, the antiparticle of the electron. This discovery opened up new avenues of research and led to the development of particle accelerators and the study of high-energy physics.

Furthermore, Dirac's work on quantum electrodynamics laid the groundwork for the development of *quantum field theory*, which is now a cornerstone of modern physics. His mathematical formalism provided a framework for understanding the behavior of particles and their interactions, leading to the development of the *Standard Model*, which describes the fundamental particles and forces in the universe. Paul Dirac's contributions to the field of physics have left an indelible mark on our understanding of the universe. His groundbreaking work in quantum mechanics and quantum electrodynamics revolutionized the field and paved the way for further advancements in theoretical physics. Dirac's legacy as a pioneer in the field will continue to inspire future generations of scientists to push the boundaries of knowledge and unravel the mysteries of the universe.

# ERWIN

# SCHRODINGER

Erwin Schrödinger, an Austrian physicist and Nobel laureate, is widely recognized for his groundbreaking contributions to the field of quantum mechanics. Born on August 12, 1887, in Vienna, Schrödinger's work revolutionized our understanding of the atomic and subatomic world. This writing aims to analyze the life and scientific contributions of Erwin Schrödinger, adopting an analytical approach.

Erwin Schrödinger's early life played a significant role in shaping his scientific career. He grew up in a family of intellectuals, with his father being a botanist and his mother coming from a family of scholars. This environment fostered his curiosity and love for knowledge from an early age. Schrödinger pursued his education at the University of Vienna, where he studied physics and mathematics under the guidance of *renowned scientists such as Friedrich Hasenöhrl and Franz Serafin Exner.*

Schrödinger's most notable contribution to science was the development of wave mechanics, a fundamental theory in quantum mechanics. In 1926, he published his groundbreaking paper titled Quantization as an Eigenvalue Problem, which laid the foundation for *wave mechanics*. This theory introduced the concept of *wave functions*, which describe the behavior of particles in terms of probabilities rather than deterministic outcomes.

One of Schrödinger's most famous achievements is the Schrödinger equation, which mathematically describes the behavior of quantum systems. This equation revolutionized the field of quantum mechanics by providing a framework to calculate the wave function of a particle at any given time. The Schrödinger equation has become a cornerstone of modern physics and has been instrumental in numerous scientific advancements.

Furthermore, Schrödinger's work on wave mechanics led to the discovery of the *wave-particle duality*, a fundamental concept in quantum physics. He demonstrated that particles, such as electrons, can exhibit both wave-like and particle-like properties, depending on the experimental setup. This duality challenged the classical understanding of physics and paved the way for further exploration into the nature of matter and energy. Erwin Schrödinger's contributions to science have had a lasting impact on our understanding of the physical world. His wave mechanics provided a more comprehensive and accurate description of atomic and subatomic phenomena, surpassing the limitations of classical physics. Schrödinger's work laid the groundwork for future developments in quantum mechanics, inspiring generations of scientists to delve deeper into the mysteries of the quantum realm.

Moreover, Schrödinger's wave equation has found practical applications in various fields, including chemistry, material science, and electronics. It has enabled scientists to predict and understand the behavior of particles and systems at the atomic level, leading to advancements in technology and the development of new materials. Erwin Schrödinger's life and scientific contributions have left an indelible mark on the field of physics. *His development of wave mechanics and the Schrödinger equation* revolutionized our understanding of quantum phenomena, challenging classical physics and opening new avenues for scientific exploration. Schrödinger's work continues to inspire and guide scientists in their quest to unravel the mysteries of the subatomic world. His legacy serves as a testament to the power of human curiosity and the pursuit of knowledge.

# NEILS BOHR

Niels Bohr, a Danish physicist, is widely recognized as one of the most influential figures in the development of quantum mechanics. His groundbreaking work on atomic structure and the understanding of the behavior of electrons revolutionized the field of physics. We will analyze the contributions of Niels Bohr to the field of quantum mechanics, focusing on his atomic model and the principle of complementarity.

One of Bohr's most significant contributions was his *atomic model*, proposed in 1913. Prior to Bohr's model, the understanding of atomic structure was limited, with scientists struggling to explain the stability of atoms and the emission of light. Bohr's model introduced the concept of quantized energy levels, where electrons occupy specific orbits around the nucleus. This model successfully explained the stability of atoms and the discrete nature of atomic spectra.

Bohr's model was based on the postulate that electrons can only exist in certain energy levels, and they transition between these levels by absorbing or emitting energy in discrete packets called quanta. This concept was a departure from classical physics, which suggested that electrons could occupy any energy level continuously. Bohr's model provided a framework for understanding the behavior of electrons and their interaction with electromagnetic radiation.

Another significant contribution of Bohr was the principle of complementarity, which he introduced in the late 1920s. This principle states that certain phenomena in quantum mechanics cannot be observed simultaneously in their entirety. For example, the wave-particle duality of light and matter cannot be fully understood by observing either the wave or particle properties alone. Instead, both aspects must be considered simultaneously, even though they may seem contradictory.

Bohr argued that the wave-particle duality is an inherent feature of quantum mechanics and that attempting to explain it solely in classical terms would lead to inconsistencies. *He emphasized the importance of accepting the limitations of human perception* and understanding that certain phenomena can only be described through complementary concepts. This principle had a profound impact on the philosophical interpretation of quantum mechanics and continues to shape the field to this day.

Niels Bohr's contributions to quantum mechanics have had a lasting impact on the field of physics. His atomic model provided a breakthrough in understanding the behavior of electrons and the stability of atoms. Additionally, his principle of *complementarity* challenged traditional notions of observation and paved the way for a deeper understanding of quantum phenomena. Bohr's work continues to inspire and guide scientists in their exploration of the quantum world.

# ARTHUR EDDINGTON

Arthur Eddington was a renowned British astrophysicist who made significant contributions to the field of theoretical physics and astronomy during the early 20th century. His groundbreaking research and innovative ideas revolutionized our understanding of the universe. We will analyze the life and work of Arthur Eddington, focusing on his contributions to astrophysics and his impact on the scientific community.

Arthur Stanley Eddington was born on December 28, 1882, in Kendal, England. He displayed exceptional academic abilities from a young age and was awarded a scholarship to study mathematics at Owens College, Manchester. Eddington's interest in physics and astronomy grew during his time at Owens College, leading him to pursue a career in astrophysics.

Eddington's most significant contribution

to astrophysics was his confirmation of Albert Einstein's theory of general relativity during the 1919 solar eclipse. This experiment, known as the [Eddington expedition](), involved observing the bending of starlight as it passed near the sun. Eddington's measurements provided the first empirical evidence supporting Einstein's theory, which revolutionized our understanding of gravity and the structure of the universe.

Furthermore, Eddington's work on stellar structure and evolution significantly advanced our knowledge of stars. He proposed the concept of the [Eddington luminosity](), which describes the maximum luminosity a star can achieve before the radiation pressure exceeds the gravitational force. This concept helped explain the stability and behavior of stars, providing a foundation for future research in stellar astrophysics.

Eddington also made important contributions to the understanding of the internal structure of stars. He developed the Eddington limit, which determines the maximum mass a star can have before it collapses under its own gravity. This limit is crucial in understanding the formation and evolution of stars, as it helps predict their lifespan and eventual fate.

Arthur Eddington's work had a profound impact on the scientific community, both during his lifetime and in the years that followed. His confirmation of Einstein's theory of general relativity elevated Einstein to the status of a scientific icon and solidified the theory as a cornerstone of modern physics. Eddington's experiments and observations provided crucial evidence that shaped the field of astrophysics and influenced subsequent research.

Eddington's contributions to stellar astrophysics also had a lasting impact. His theories and concepts laid the groundwork for future studies on stellar evolution and the behavior of stars. Many of his ideas are still widely used in contemporary astrophysics, demonstrating the enduring significance of his work. Arthur Eddington's pioneering work in astrophysics revolutionized our understanding of the universe. *His confirmation of Einstein's theory of general relativity and his contributions to stellar astrophysics* have had a lasting impact on the scientific community. Eddington's innovative ideas and meticulous observations continue to shape our understanding of the cosmos, making him a true pioneer in the field of astrophysics.

# GEORGE LEMAITRE

George Lemaître, a Belgian Catholic priest and physicist, is widely recognized as *the pioneer of the Big Bang theory*. His groundbreaking work in cosmology revolutionized our understanding of the universe's origins and laid the foundation for modern astrophysics. We will analyze Lemaître's contributions to the field, exploring his scientific achievements, the impact of his theory, and the academic legacy he left behind.

Lemaître's most significant scientific achievement was his proposal of <u>the Big Bang theory</u>, which he first presented in 1927. He theorized that the universe originated from a singular point, a "primeval atom," which subsequently expanded and evolved into the vast cosmos we observe today. Lemaître's theory was based on Einstein's general theory of relativity and the observed redshift of distant galaxies, which indicated that the universe was expanding. Lemaître's work was not limited to theoretical calculations; he also made significant contributions to observational astronomy. In 1929, he discovered the relationship between the distance and recession velocity of galaxies, now known as <u>Hubble's Law</u>. This discovery provided empirical evidence for the expansion of the universe and further supported Lemaître's Big Bang theory.

Lemaître's Big Bang theory revolutionized our understanding of the universe's origins and challenged prevailing notions of a static and eternal cosmos. His theory provided a scientific explanation for the observed expansion of the universe and the abundance of light elements, such as hydrogen and helium. Moreover, it offered a framework for understanding the formation of galaxies, stars, and other celestial structures. The acceptance of the Big Bang theory had profound implications for various scientific disciplines. It provided a basis for the study of cosmology, astrophysics, and the evolution of the universe. Lemaître's theory also paved the way for the development of the *cosmic microwave background radiation*, which further confirmed the validity of the Big Bang model.

Lemaître's contributions to cosmology and astrophysics were not limited to his scientific achievements. As an academic, he played a crucial role in disseminating knowledge and fostering scientific inquiry. Lemaître's work inspired subsequent generations of scientists to explore the mysteries of the universe and expand upon his theories. Furthermore, Lemaître's academic legacy extends beyond his scientific contributions. As a

Catholic priest, he sought to *reconcile science and religion*, emphasizing that *the Big Bang theory did not contradict the existence of a divine creator*. Lemaître's efforts to bridge the gap between science and religion have had a lasting impact on the dialogue between these two realms.

George Lemaître's pioneering work in cosmology and his proposal of the Big Bang theory have forever changed our understanding of the universe. His scientific achievements, including the discovery of Hubble's Law and the formulation of the Big Bang theory, have had a profound impact on the field of astrophysics. Lemaître's academic legacy extends beyond his scientific contributions, as he sought to reconcile science and religion, fostering a dialogue that continues to this day. As we continue to explore the mysteries of the cosmos, Lemaître's legacy serves as a reminder of the power of human curiosity and the potential for scientific breakthroughs.

# RICHARD

# FEYNMAN

Richard Feynman, born on May 11, 1918, was an American theoretical physicist who made significant contributions to the field of quantum mechanics and particle physics. His exceptional intellect and unique approach to problem-solving have made him one of the most influential scientists of the 20th century. We will analyze the life and work of Richard Feynman, highlighting his contributions to physics and his impact on the scientific community. Feynman's academic journey began at the Massachusetts Institute of Technology (MIT), where he obtained his bachelor's degree in physics in 1939. He then pursued his doctoral studies at Princeton University, where he worked under the guidance of renowned *physicist John Archibald Wheeler*. During this time, Feynman developed a deep understanding of quantum mechanics, a field that would become the cornerstone of his future research.

One of Feynman's most significant contributions to physics was his development of <u>the Feynman diagrams</u>. These diagrams provided a visual representation of the interactions between elementary particles, making complex calculations more accessible and intuitive. Feynman's diagrams revolutionized the field of particle physics and played a crucial role in the development of the quantum electrodynamics (QED) theory.

In addition to his theoretical work, Feynman was also an exceptional teacher and communicator. He had a unique ability to explain complex scientific concepts in a simple and engaging manner. Feynman's lectures at the California Institute of Technology (Caltech) became legendary, attracting students and researchers from various disciplines. His lectures on physics, later compiled into a series of books known as The Feynman Lectures on Physics, continue to be widely read and admired by both students and professionals.

Feynman's passion for science extended beyond the confines of academia. He actively participated in the Manhattan Project during World War II, contributing to the development of the atomic bomb. However, Feynman's involvement in the project did not deter him from questioning the ethical implications of his work. He later became an advocate for scientific integrity and emphasized the importance of responsible scientific research. Furthermore, Feynman's curiosity and love for adventure led him to explore various fields outside of physics. He had a keen interest in biology, studying the behavior of ants and investigating the process of DNA replication. Feynman's interdisciplinary approach to science allowed him to make connections between different disciplines, leading to new insights and discoveries.

Richard Feynman's impact on the scientific community cannot be overstated. His contributions to quantum mechanics and particle physics have shaped our understanding of the fundamental laws of nature. Moreover, his dedication to teaching and his ability to communicate complex ideas have inspired countless students and researchers to pursue careers in science. Richard Feynman was a brilliant physicist whose contributions to the field of physics continue to resonate today. His development of the Feynman diagrams revolutionized the way we understand particle interactions, while his engaging teaching style made science accessible to a broader audience. Feynman's interdisciplinary approach and his commitment to scientific integrity serve as an inspiration to future generations of scientists. His legacy as a brilliant mind and a passionate educator will forever be remembered in the annals of scientific history.

# WOLFGANG

# PAULI

Wolfgang Pauli, born on April 25, 1900, in Vienna, Austria, was a renowned physicist who made significant contributions to the field of quantum mechanics. His work revolutionized our understanding of the atomic and subatomic world, and his ideas continue to shape the foundations of modern physics. We will analyze the life and achievements of Wolfgang Pauli, focusing on his contributions to quantum mechanics and his impact on the scientific community.

Pauli's interest in physics emerged at an early age, and he pursued his passion by enrolling at the University of Munich in 1918. Under the guidance of *Arnold Sommerfeld*, a prominent physicist of the time, Pauli developed a deep understanding of theoretical physics. He completed his doctoral thesis in 1921, which focused on <u>the theory of relativity</u>, a topic that would later influence his groundbreaking work in quantum mechanics.

Pauli's most significant contribution to quantum mechanics was his formulation of <u>the Pauli exclusion principle in 1925</u>. This principle states that *no two identical fermions*, such as electrons, *can occupy the same quantum state simultaneously*. This discovery had profound implications for our understanding of the behavior of electrons in atoms and *laid the foundation for the development of the periodic table of elements*. Furthermore, Pauli's work on quantum mechanics extended beyond the exclusion principle. He made significant contributions to <u>the theory of spin</u>, proposing the concept of spin angular momentum in 1927. This concept explained the magnetic properties of atoms and provided a deeper understanding of the behavior of particles at the atomic level.

Pauli's contributions to quantum mechanics had a profound impact on the scientific community. His exclusion principle provided a theoretical framework for understanding the stability and electronic structure of atoms. It also paved the way for the development of quantum field theory, which has become a cornerstone of modern physics. Moreover, Pauli's work influenced a generation of physicists, including his close collaborator, Werner Heisenberg, and his student, *Carl Friedrich von Weizsäcker*. His ideas sparked further research and led to the development of new theories and experimental techniques. Pauli's influence can still be seen in the work of contemporary physicists, who continue to build upon his foundational contributions.

Wolfgang Pauli's contributions to quantum mechanics have solidified his place as one of the most influential physicists of the 20th century. His formulation of <u>the exclusion principle and his work on spin</u> revolutionized our understanding of the atomic and subatomic world. Pauli's impact on the scientific community cannot be overstated, as his ideas continue to shape the foundations of modern physics. Through his analytical approach and academic rigor, Pauli left an indelible mark on the field of quantum mechanics, and his legacy will continue to inspire future generations of physicists.

# CARL SAGAN

Carl Sagan, an eminent astronomer, astrophysicist, and science communicator, left an indelible mark on popularizing science and inspiring awe and curiosity about the cosmos. His profound impact transcends the realms of scientific research, encompassing an enduring legacy as a visionary advocate for scientific literacy, critical thinking, and environmental stewardship. Sagan's unparalleled ability to communicate the wonders of the universe to a global audience through his bestselling books and seminal television series, such as <u>Cosmos: A Personal Voyage</u>, and captivating lectures, galvanized an unprecedented enthusiasm for astronomy, planetary science, and the exploration of the universe.

His pioneering contributions to the scientific understanding of planetary atmospheres and the search for extraterrestrial life were marked by his involvement in NASA's Mariner, Viking, and Voyager missions, which provided unprecedented insights into the properties of other planets within our solar system and the prospects for life beyond Earth. Additionally, Sagan's instrumental role in the scientific community extended to his advocacy for nuclear disarmament, the mitigation of global environmental challenges and the promotion of critical thinking and evidence-based reasoning in addressing complex societal issues. His steadfast commitment to addressing these pressing global concerns underscored the profound impact science and humanistic principles can have in shaping a more informed, compassionate, and sustainable world. Sagan's legacy continues to resound via his evocative writings, fervent advocacy for scientific discovery, and his unrelenting dedication to cultivating a better appreciation for the exquisite beauty and profound interconnection of the cosmos.

# ARTHUR C.

# CLARK

Arthur C. Clarke was a visionary science-fiction writer, futurist, and inventor, who envisioned and popularized concepts that have now become important elements of modern technology breakthroughs. His imaginative storytelling prowess, coupled with his prescient insights into the potential of science and technology, established him as a seminal figure in the realms of science fiction literature and technological prognostication Clarke's most renowned work, <u>2001: A Space Odyssey</u>, exemplifies his ability to craft compelling narratives that seamlessly intertwine speculative fiction with plausible scientific concepts. The exploration of advanced artificial intelligence, interplanetary travel, and enigmatic encounters with extraterrestrial intelligence within the narrative of <u>2001: A Space Odyssey</u>.

Foreshadowed technical breakthroughs and existential issues that firmly resound within modern scientific and philosophical discourse. Furthermore, Clarke's visionary writings and prognostications stretched beyond the world of fiction, as he posited technological concepts, such the geostationary communications satellite, in a manner that proved essential in defining the course of later technological breakthroughs.

His visionary articulation of satellite telecommunications technology profoundly impacted the eventual realization and widespread adoption of this pivotal technological innovation, amplifying the interconnectedness and capabilities of modern global communication networks. Moreover, Clarke's boundless enthusiasm and insightful predictions regarding humanity's potential for space exploration and the prospects for extraterrestrial encounters continue to inspire and inform contemporary scientific and technological pursuits.

His visionary expositions on space elevators, lunar infrastructure, and the imperative of venturing beyond the confines of Earth have stimulated enduring interest and research endeavors within the realms of aerospace engineering planetary science, and space exploration. In commemorating Arthur C. Clarke's indelible influence on science fiction, technology and the envisioning of humanity's future, we acknowledge his legacy as a harbinger of technological innovation, scientific inquiry, and the enduring human quest for understanding, exploration, and transcendence. Clarke's imaginative sagacity and his capacity to envisage technology's transformative impact on the human condition continue to inspire innovators, storytellers, and futurists in their pursuit of advancing the frontiers of knowledge, imagination, and societal progress.

# STEPHEN

# HAWKING

Stephen Hawking was a great theoretical physicist, cosmologist, and novelist whose innovative research and engaging popularization of science altered our knowledge of the world and its fundamental principles. Despite battling amyotrophic lateral sclerosis (ALS), Hawking's intellectual agility, relentless curiosity, and innovative contributions propelled him to the vanguard of theoretical physics, establishing him as an iconic figure revered for his indomitable spirit and unparalleled scientific acumen. Hawking's seminal work on black holes (particularly) and his groundbreaking theoretical insights into their thermodynamic properties and the emission of <u>Hawking radiation</u>, transformed our comprehension of these enigmatic cosmic entities, reshaping the theoretical landscape of general relativity quantum mechanics, and the fundamental nature of space, time, and gravity.

His bestselling book, <u>A Brief History of Time</u>, captivated audiences across the globe with its lucid explications of complex astrophysical and cosmological concepts, rendering intricate scientific principles accessible to a broad readership and igniting a resurgent interest in the mysteries of the cosmos.

Throughout his unparalleled career, Hawking's interdisciplinary explorations extended to cosmological inquiries, quantum gravity, the origins of the universe, and the reconciliation of quantum mechanics with general relativity exemplifying his relentless pursuit of a unified framework that encapsulates the inherent dynamics of the cosmos. Beyond his astrophysical prowess, Hawking's resolute advocacy for science education, his steadfast commitment to demystifying complex scientific concepts, and his engagement with broader societal issues underscored his enduring impact as an erudite public intellectual and an exemplar of scientific inquiry and outreach.

In commemorating Stephen Hawking's profound contributions to theoretical physics, cosmology, and science communication, we honor his enduring legacy as a paragon of scientific investigation, intellectual fortitude and the unyielding pursuit of knowledge. His resounding impact on the public's understanding of the universe, coupled with his steadfast determination to surmount physical adversities, reinforces his stature as an emblem of resilience, intellect, and empathetic engagement, perpetuating his legacy as an enduring source of inspiration for scientists, educators, and enthusiasts alike.

MICHIO KAKU

Michio Kaku is a renowned theoretical physicist, futurist, and popular science communicator. Born on January 24, 1947, in San Jose, California, Kaku has made significant contributions to the field of theoretical physics, particularly in the study of *string theory*. This writing will analyze Kaku's contributions to the field, his impact on popularizing science, and his role as a futurist.

One of Michio Kaku's most notable contributions to theoretical physics is his work on string theory. String theory is a branch of theoretical physics that attempts to explain the fundamental nature of particles and the forces that govern them. Kaku has been instrumental in developing and popularizing this theory, which posits that *particles are not point-like entities but rather tiny vibrating strings.*

Kaku's research has focused on the mathematical aspects of string theory, exploring its potential to unify the four fundamental forces of nature: gravity, electromagnetism, the strong nuclear force, and the weak nuclear force. His work has shed light on the possibility of a grand unified theory, which could provide a comprehensive understanding of the universe at its most fundamental level. In addition to his contributions to string theory, Kaku has also made significant advancements in the field of quantum mechanics. He has explored the concept of quantum field theory, which describes the behavior of particles and fields at the quantum level. Kaku's research has helped to deepen our understanding of the strange and counterintuitive phenomena that occur at this scale, such as quantum entanglement and superposition.

Beyond his contributions to theoretical physics, Michio Kaku has played a vital role in popularizing science and making complicated concepts accessible to the general public. Through his numerous books, television appearances, and public lectures, Kaku has brought scientific ideas to a large audience. Kaku's ability to explain complex scientific concepts in a clear and engaging manner has made him a beloved figure in the scientific community and beyond. He has a talent for breaking down complex theories into relatable analogies, allowing non-experts to grasp the fundamental ideas behind them. This skill has made him a sought-after guest on television shows and a popular speaker at conferences and events.

As a science communicator, Kaku has also emphasized the importance of inspiring the next generation of scientists. He has advocated for increased investment in science education and has worked to make science more accessible and exciting for young people. Through his efforts, Kaku has helped to foster a greater interest in science among students and the general public.

In addition to his work in theoretical physics and science communication, Michio Kaku is noted for his role as a futurist. He has written extensively about the future of technology, examining issues such as artificial intelligence, nanotechnology, and space exploration. Kaku's futurist perspectives are grounded in his scientific background, allowing him to provide informed insights into the potential advancements and challenges that lie ahead. He has discussed the possibilities of colonizing other planets, achieving immortality through advancements in medicine, and harnessing the power of artificial intelligence for the betterment of society.

While some critics argue that Kaku's futurist predictions may be overly optimistic or speculative, his ideas have sparked important discussions about the ethical and societal implications of emerging technologies. By encouraging dialogue and debate, Kaku has helped to shape public discourse on the future of science and technology.

Michio Kaku's contributions to theoretical physics, his efforts in popularizing science, and his role as a futurist have made him a prominent figure in the scientific community. Through his research, Kaku has advanced our understanding of the fundamental nature of the universe. As a science communicator, he has made complex scientific concepts accessible to a wide audience, inspiring a greater interest in science. Finally, as a futurist, Kaku has sparked important discussions about the potential impact of emerging technologies on society. Overall, Michio Kaku's work has had a profound influence on the field of theoretical physics and the public's perception of science.

# ALAN GUTH

Alan Guth, an American theoretical physicist, is widely recognized for his groundbreaking contributions to the field of cosmology. His work on the theory of cosmic inflation revolutionized our understanding of the early universe and provided a solution to several long-standing problems in cosmology. This writing aims to analyze Guth's contributions to the field of cosmology, focusing on his development of the inflationary theory, its implications, and its impact on our understanding of the universe. In the early 1980s, Guth proposed the concept of *cosmic inflation*, a theory that explains the rapid expansion of the universe in its earliest moments. Prior to Guth's work, the Big Bang theory was the prevailing explanation for the origin of the universe. However, this theory faced several challenges, such as the horizon problem and the flatness problem.

*The horizon problem* arises from the fact that regions of the universe that are far apart appear to have *the same temperature*, despite not having enough time to exchange energy or information. *The flatness problem*, on the other hand, questions why the universe appears to be so close to flatness, which is a critical condition for the formation of galaxies and other structures. Guth's inflationary theory proposes that the universe underwent a brief period of *exponential expansion*, driven by a hypothetical field called *the inflation*. This rapid expansion would have smoothed out the irregularities in the early universe, explaining the uniformity observed in the cosmic microwave background radiation. Additionally, the inflationary theory predicts that the universe should be nearly flat, providing a solution to the flatness problem.

Guth's inflationary theory has had profound implications for our understanding of the universe. It not only addresses the horizon and flatness problems but also provides an explanation for the origin of the large-scale structure of the universe, such as galaxies and galaxy clusters. The theory suggests that quantum fluctuations during inflation could have seeded the density variations that eventually led to the formation of these structures. Furthermore, the inflationary theory predicts the existence of gravitational waves, which are ripples in the fabric of spacetime. These gravitational waves were recently detected by the BICEP2 experiment, providing strong evidence in support of Guth's theory. *The discovery of these primordial gravitational waves* has opened up new avenues for studying the early universe and has garnered significant attention from the scientific community.

Guth's work has not only advanced our understanding of the universe but has also inspired further research and theoretical developments. Many variations of the inflationary theory have been proposed, each with its own set of predictions and implications. These ongoing investigations continue to shape our understanding of the early universe and its evolution. Alan Guth's contributions to cosmology, particularly his development of <u>the inflationary theory</u>, have had a profound impact on our understanding of the universe. His work has provided solutions to long-standing problems in cosmology and has opened up new avenues for research. Guth's pioneering efforts have not only advanced our knowledge of the early universe but have also inspired subsequent generations of physicists to explore the mysteries of the cosmos.

# HISTORICAL BLACK SCIENTISTS

# BENJAMIN BANNEKER

Benjamin Banneker was an extraordinary figure in the history of the United States. Born in 1731 to a free African American family in Maryland, Banneker went on to become a *self-taught mathematician, astronomer, and inventor*. Despite facing racial discrimination, he became a prominent scientist, writer, and surveyor. He was a Renaissance man in every sense of the word. Banneker's legacy is a testament to the power of knowledge and perseverance. His contributions to science, literature, and civil rights continue to inspire generations of people around the world. In this post, we will explore the remarkable life and work of Benjamin Banneker and how his achievements paved the way for future generations.

Benjamin Banneker, a name that echoes through the corridors of history as a symbol of brilliance, resilience, and intellectual prowess. Born in 1731 in Maryland, Banneker was an extraordinary African American man who defied the odds and left an indelible mark on the world. In an era plagued by racial prejudice and inequality, Banneker emerged as a shining example of what one can achieve with determination, intellect, and an insatiable thirst for knowledge.

Banneker's significance lies not only in his remarkable accomplishments but also in the context of the time in which he lived. The 18th century was a period of great societal and intellectual transformation, commonly known as the Renaissance. It was during this time that Banneker's genius flourished, and he became a true Renaissance man.

While primarily known as an astronomer

and mathematician, Banneker's talents spanned a wide range of disciplines. He was not only an accomplished inventor but also an expert in agriculture, surveying, and literature. His love for learning knew no bounds, and he dedicated his life to expanding his knowledge and exploring various fields of study.

Banneker's most notable achievement came in the field of astronomy. Armed with only a borrowed pocket watch and self-taught knowledge, *he accurately predicted celestial events and meticulously calculated ephemerides.* His <u>astronomical almanacs</u>, published annually, became invaluable tools for farmers, sailors, and scholars alike. Beyond his intellectual pursuits, Banneker was a staunch advocate for civil rights and equality. He corresponded with prominent figures of his time, including Thomas Jefferson, challenging the prevailing notion of racial inferiority and calling for the abolition of slavery. His eloquent words and unwavering commitment to justice continue to inspire generations to fight for equality.

Born on November 9, 1731, in Ellicott's Mills, Maryland, Benjamin Banneker was the son of Robert and Mary Banneker, both descendants of enslaved Africans. Despite being born into a world plagued by slavery and racial inequality, Banneker's parents instilled in him the values of education, hard work, and intellectual curiosity. Banneker's formal education was limited, as he only attended a small Quaker school for a few years. However, his thirst for knowledge knew no bounds. He became an avid reader, devouring every book he could get his hands on, and developed a keen interest in mathematics and astronomy.

Living on his family's farm, Banneker quickly showcased his intellectual prowess by building *a functioning wooden clock* entirely from scratch at the young age of 22. This remarkable achievement not only demonstrated his ingenuity but also marked the beginning of his lifelong dedication to scientific pursuits. Despite the limited opportunities available to African Americans during that era, Banneker's intellect and skills attracted the attention of notable figures, including George Ellicott, who recognized his incredible potential. Ellicott provided Banneker with access to books, instruments, and mentorship, nurturing his burgeoning talent.

Banneker's insatiable thirst for knowledge led him to become *a self-taught mathematician, astronomer, surveyor, and writer.* His groundbreaking almanacs, which he meticulously calculated and published annually, provided valuable information on astronomy, tides, weather patterns, and agricultural advice. These almanacs not only gained him widespread recognition but also challenged the prevailing notions of African Americans' intellectual capabilities. Benjamin Banneker's early life and upbringing were marked by a relentless pursuit of knowledge, fueled by the unwavering support of his family and mentors. His remarkable achievements in various fields would serve as an inspiration for generations to come, proving that brilliance knows no racial or societal barriers. The legacy of this Renaissance man continues to shine brightly, reminding us of the transformative power of education, passion, and determination. Benjamin Banneker's contributions to astronomy and mathematics are truly remarkable and have left an indelible mark on history. Despite being largely self-taught, Banneker's knowledge in these fields was vast and groundbreaking.

In astronomy, Banneker's achievements were nothing short of extraordinary. He meticulously studied celestial bodies and accurately calculated their movements and positions. His most notable accomplishment was the creation of a series of almanacs, which provided *detailed astronomical and meteorological data*. These almanacs were not only widely acclaimed but also served as important tools for farmers and navigators. Banneker's mathematical prowess was equally impressive. He excelled in complex calculations and geometric principles, often solving intricate problems with ease. His *mastery of mathematics* greatly influenced his work in astronomy, allowing him to make precise measurements and predictions.

One of Banneker's most significant contributions was *his involvement in surveying the land for the construction of Washington, D.C.* His mathematical and astronomical expertise were crucial in accurately laying out the city's streets and boundaries. His meticulous calculations and attention to detail ensured the success of this monumental project. Banneker's accomplishments in astronomy and mathematics challenged the prevailing notion that African Americans were intellectually inferior. He shattered stereotypes and paved the way for future generations of African American scientists and mathematicians.

Today, Banneker's legacy continues to inspire and captivate. His contributions to astronomy and mathematics serve as a testament to the power of knowledge, determination, and a relentless pursuit of excellence. Benjamin Banneker's remarkable achievements will forever be celebrated as an integral part of our shared intellectual heritage. Benjamin Banneker's remarkable contributions extend beyond his achievements as an astronomer and mathematician. His role in *surveying and the creation of Washington, D.C.* showcases his versatility and ingenuity in shaping the landscape of the nation's capital.

In the late 18th century, the United States faced the monumental task of designing and building a new capital city. Banneker's involvement in this process was nothing short of extraordinary. As one of the few African Americans involved in such a monumental project, his contributions were groundbreaking in more ways than one. Banneker's expertise in surveying played a pivotal role in laying out the plans for the city. Working alongside Pierre Charles L'Enfant, the appointed architect for the city, Banneker meticulously surveyed the land, ensuring accurate measurements and precise boundaries. His keen eye for detail and mathematical aptitude ensured that the city's layout would be both functional and aesthetically pleasing.

Through his correspondence with prominent figures such as Thomas Jefferson, Banneker fearlessly addressed issues of slavery, racial inequality, and the importance of recognizing the humanity and intellect of African Americans. His words and actions served as a catalyst for change and challenged the prevailing beliefs of his time.

Banneker's commitment to education and advocacy continues to inspire generations. His legacy reminds us of the transformative power of knowledge and the importance of fighting for equal opportunities for all. Benjamin Banneker's remarkable achievements as a Renaissance man are undoubtedly impressive, but his dedication to education and advocacy for African Americans truly sets him apart as an extraordinary figure in history. Banneker's genius extended far beyond his mathematical and scientific pursuits. He was also an avid writer and engaged in correspondence with many notable figures of his time. Through his eloquent and insightful letters, Banneker showcased his intellect, passion, and unwavering dedication to the causes he believed in.

One of his most famous correspondences

was with Thomas Jefferson, then Secretary of State, where Banneker implored him to reconsider his stance on slavery and the inherent contradiction between the ideals of liberty and the practice of owning enslaved individuals. In his letter, Banneker eloquently argued for the equality and humanity of all people, irrespective of their race.

Banneker's writings were not limited to political matters. He also documented his astronomical observations and calculations, providing valuable insights into the celestial events of his time. His almanacs, filled with weather predictions, astronomical data, and practical advice, were widely read and respected. Moreover, Banneker's writings extended to poetry and literature, showcasing his artistic and creative side. His poem, To the Public, is a testament to his poetic talents and his desire to use his gifts to uplift and inspire others.

In addition to his mechanical prowess, Banneker's astronomical observations and calculations were groundbreaking. He accurately predicted celestial events such as solar and lunar eclipses, displaying a level of astronomical expertise that surpassed many of his contemporaries. His almanacs, which included tide tables, weather predictions, and farming advice, became invaluable resources for farmers and navigators alike. Though Banneker's life was tragically cut short, his impact continues to inspire generations. His relentless pursuit of knowledge, his commitment to justice, and his remarkable achievements in various disciplines serve as a testament to the power of intellect, determination, and resilience. Benjamin

Banneker's legacy is a shining example of what can be accomplished against all odds and remains an enduring inspiration for those who dare to dream and strive for greatness.

Throughout his remarkable life, Benjamin Banneker received numerous honors and recognition for his exceptional contributions to various fields. Despite facing adversity and discrimination due to his race, Banneker's brilliance and perseverance earned him widespread admiration and respect. One of the earliest honors came in the form of a letter from Thomas Jefferson, who was then serving as the Secretary of State. Impressed by Banneker's extraordinary mathematical abilities, Jefferson wrote a letter in 1791 expressing his astonishment and acknowledging Banneker's intellect. This correspondence sparked a brief friendship and a dialogue on the topic of racial equality. Banneker's achievements in astronomy were also highly regarded. His meticulous calculations and predictions regarding celestial events, such as his accurate projection of a solar eclipse in 1789, earned him respect among fellow astronomers and scientists of his time. His almanacs, which contained essential information such as tide tables, weather forecasts, and astronomical data, were widely praised for their accuracy and usefulness.

The abolitionist movement recognized Banneker as an influential figure in the fight against slavery. He corresponded with prominent abolitionists, including Benjamin Rush and Thomas Jefferson, advocating for the abolition of slavery and the equal treatment of all individuals. Banneker's eloquent arguments and logical reasoning greatly contributed to the intellectual and moral foundation of the abolitionist cause. In recognition of his accomplishments, Banneker was appointed by President George Washington to serve on a commission responsible for the surveying and planning of the nation's capital, Washington, D.C. Though he faced prejudice and discrimination from some of his fellow commissioners, Banneker's expertise and dedication to the project were undeniable. Today, Benjamin Banneker's legacy continues to be honored and celebrated. Numerous schools, buildings, and institutions bear his name, serving as a testament to his enduring impact. His contributions to mathematics, astronomy, civil rights, and the advancement of knowledge serve as an inspiration to generations, reminding us of the remarkable achievements that can be accomplished in the face of adversity.

Furthermore, Banneker's dedication to justice and equality is a lesson that resonates deeply. He used his knowledge and skills to fight against the injustices faced by African Americans, advocating for their rights and challenging the prevailing societal norms of the time. Banneker's *activism* reminds us of the importance of *using our talents and platforms to stand up for what is right, even in the face of opposition.* His legacy inspires us to be agents of change and to work towards creating a more just and inclusive society. The life and legacy of Benjamin Banneker are truly remarkable. As a self-taught mathematician, astronomer, inventor, and writer, Banneker defied the societal limitations placed upon African Americans during his time and left an indelible mark on history.

# NEIL DEGRASSE

# TYSON

Neil deGrasse Tyson is a renowned astrophysicist, science communicator, and public intellectual who has made significant contributions to the field of astronomy. With his charismatic personality and ability to explain complex scientific concepts in a relatable manner, Tyson has become a household name and an inspiration to many aspiring scientists. This chapter will explore the life and achievements of Neil deGrasse Tyson, highlighting his impact on the scientific community and society as a whole.

Born on October 5, 1958, in New York City, Tyson developed a passion for astronomy at a young age. His interest was sparked during a visit to the Hayden Planetarium, where he was captivated by the wonders of the universe. This experience set him on a path towards a career in astrophysics, eventually leading him to earn a Bachelor of Arts in Physics from Harvard University and a Ph.D. in Astrophysics from Columbia University.

Tyson's contributions to the scientific community are vast and varied. He has conducted extensive research on star formation, galactic evolution, and the structure of our Milky Way galaxy. His work has been published in numerous scientific journals, earning him recognition and respect among his peers. Additionally, Tyson has served as the director of the Hayden Planetarium since 1996, where he has played a crucial role in promoting public understanding of science.

One of Tyson's most significant achievements is his ability to communicate complex scientific ideas to the general public. Through his books, television shows, and public appearances, he has made science accessible and engaging for people of all ages and backgrounds. Tyson's unique talent lies in his ability to break down complex concepts into simple, relatable terms, making science more approachable for those who may have previously found it intimidating.

Tyson's television series, such as <u>Cosmos: A Spacetime Odyssey</u> and <u>Star-Talk</u>, have reached millions of viewers worldwide. These shows not only educate the public about scientific discoveries and theories but also inspire a sense of wonder and curiosity about the universe. Tyson's charismatic personality and enthusiasm for his subject matter make him a captivating presenter, drawing in audiences and igniting a passion for science in the hearts of many.

In addition to his scientific contributions, Tyson has also been an advocate for diversity and inclusion in the scientific community. He has spoken out about the importance of providing equal opportunities for underrepresented groups in STEM fields, encouraging young people from all backgrounds to pursue careers in science. Tyson's efforts to promote diversity and inclusivity have helped to break down barriers and create a more inclusive scientific community.

Neil deGrasse Tyson's impact on the scientific community and society as a whole cannot be overstated. Through his research, science communication, and advocacy, he has inspired countless individuals to explore the wonders of the universe and pursue careers in science. Tyson's ability to make complex scientific concepts accessible and relatable has made him a beloved figure in popular culture. His contributions to the field of astrophysics and his dedication to promoting diversity and inclusion have solidified his status as a scientific icon.

# KATHERINE JOHNSON

Kathleen Johnson, also known as Katherine G. Johnson, was an African-American mathematician and aerospace engineer who made significant contributions to the field of space exploration. Her work at NASA during the Space Race era played a crucial role in the success of numerous missions, including the historic Apollo 11 moon landing. We will analyze Kathleen Johnson's life, her contributions to aerospace engineering, and the impact she had on breaking barriers for women and minorities in STEM fields.

Kathleen Johnson was born on August 26, 1918, in White Sulphur Springs, West Virginia. Despite growing up in a racially segregated society, she displayed exceptional mathematical abilities from a young age. Recognizing her talent, her parents encouraged her to pursue her education, which was a rare opportunity for African-

American children at the time. Johnson attended West Virginia State College, where she graduated summa cum laude with degrees in mathematics and French in 1937. After graduation, she began teaching at a segregated public school in Marion, Virginia. However, her passion for mathematics led her to apply for a position at the National Advisory Committee for Aeronautics (NACA), which later became NASA.

In 1953, Johnson joined the all-black West Area Computing section at NACA's Langley Research Center. Despite facing racial and gender discrimination, she quickly proved her exceptional skills in analytical geometry and trajectory analysis. Her calculations were instrumental in the success of several high-profile projects, including the Mercury and Apollo missions.

One of Johnson's most notable contributions was her work on the trajectory calculations for the first American human spaceflight, Alan Shepard's Freedom 7 mission in 1961. Her precise calculations ensured that Shepard's spacecraft would return safely to Earth. Johnson's expertise in celestial mechanics also played a crucial role in the Apollo program, where she calculated the trajectories for the lunar lander and the command module. Furthermore, Johnson's calculations were pivotal in the success of the Apollo 11 mission, which resulted in the first human landing on the moon. Her work on the Lunar Module Descent Trajectory ensured that Neil Armstrong and Buzz Aldrin could safely land on the lunar surface and return to the command module for their journey back to Earth.

Kathleen Johnson's achievements were not

only groundbreaking in terms of scientific advancements but also in breaking barriers for women and minorities in STEM fields. As an African-American woman, she faced numerous challenges and prejudices throughout her career. However, her perseverance and exceptional skills shattered stereotypes and paved the way for future generations of women and minorities in the aerospace industry. Johnson's services were acknowledged with various prizes and medals, including the Presidential Medal of Freedom, which she earned in 2015. Her story was also popularized in the book and subsequent film adaptation, "Hidden Figures," which exposed her achievements to a wider audience.

Kathleen Johnson's remarkable career as a mathematician and aerospace engineer at NASA has left an indelible mark on the field of space exploration. Her groundbreaking calculations and trajectory analysis were instrumental in the success of numerous missions, including the historic Apollo 11 moon landing. Moreover, her perseverance and determination in the face of adversity have inspired generations of women and minorities to pursue careers in STEM fields. Kathleen Johnson's legacy will forever be remembered as a trailblazer who defied societal barriers and contributed significantly to the advancement of human knowledge and exploration.

# HAKEEM M. OLUSEYI

Hakeem M. Oluseyi is a renowned astrophysicist, inventor, and educator who has made significant contributions to the field of science. His work has not only advanced our understanding of the universe but has also inspired countless individuals to pursue careers in STEM fields. This writing will analyze the achievements and impact of Hakeem M. Oluseyi, taking an academic and analytical approach. Oluseyi's academic journey began at Tougaloo College, where he earned a Bachelor of Science degree in Physics. He then pursued graduate studies at Stanford University, where he obtained a Ph.D. in Physics. This solid educational foundation laid the groundwork for his subsequent groundbreaking research and inventions.

One of Oluseyi's notable contributions to astrophysics is his work on the development of novel instruments for space exploration. He has been involved in *the design and construction of advanced detectors and sensors* that have been deployed on various NASA missions. These instruments have enabled scientists to gather crucial data about celestial bodies, such as stars and galaxies, leading to significant discoveries and a deeper understanding of the universe.

In addition to his technical contributions, Oluseyi has also made significant strides in science education and outreach. He firmly believes in the importance of making science accessible to all, particularly underrepresented communities. As a professor at the Florida Institute of Technology, he has mentored numerous students, encouraging them to pursue careers in STEM fields. Oluseyi has also been actively involved in outreach programs, delivering engaging talks and workshops to inspire young minds and foster a love for science.

Oluseyi's impact extends beyond academia and research. He has appeared on various television programs, including the National Geographic Channel's <u>How the Universe Works</u> and the Science Channel's <u>Outrageous Acts of Science</u>. Through these platforms, he has been able to reach a wider audience, sharing his knowledge and passion for astrophysics. By making complex scientific concepts accessible and relatable, Oluseyi has played a crucial role in popularizing science and inspiring a new generation of scientists. Furthermore, Oluseyi's personal journey is a testament to the power of perseverance and determination. Growing up in a disadvantaged neighborhood, he faced numerous challenges and setbacks. However, he overcame these obstacles through his unwavering dedication to education and his passion for science. His story serves as an inspiration to individuals from similar backgrounds, proving that with hard work and resilience, one can achieve greatness.

Hakeem M. Oluseyi's contributions to astrophysics, science education, and outreach have had a profound impact on the scientific community and society as a whole. His work in developing advanced instruments for space exploration has expanded our knowledge of the universe, while his efforts in science education have inspired countless individuals to pursue careers in STEM fields. Through his television appearances and personal journey, Oluseyi has made science accessible and relatable to a wider audience. Hakeem M. Oluseyi's achievements and impact are a testament to his brilliance, dedication, and commitment to advancing scientific knowledge.

# DOROTHY

# VAUGHN

Dorothy Vaughan, an African-American mathematician and computer programmer, played a pivotal role in the advancement of space exploration during the early years of NASA. Her exceptional skills, determination, and leadership abilities not only broke barriers for women and minorities but also significantly contributed to the success of the United States' space program. This chapter aims to analyze the remarkable contributions of Dorothy Vaughan, highlighting her impact on the field of mathematics and her role in promoting diversity and equality within the scientific community.

Dorothy Vaughan's journey began in the 1940s when she joined the National Advisory Committee for Aeronautics (NACA), which later became NASA. As a "human computer," Vaughan was responsible for performing complex calculations by hand, a task that required exceptional mathematical skills and attention to detail. Her proficiency in mathematics and her ability to solve intricate problems quickly earned her recognition among her peers.

One of Vaughan's most significant contributions was her role in the West Area Computing Unit, where she became the first African-American supervisor. In this position, she not only managed a team of talented mathematicians but also advocated for equal opportunities for women and minorities. Vaughan's leadership skills and dedication to promoting diversity within the scientific community were instrumental in breaking down racial and gender barriers. Furthermore, Vaughan's expertise in computer programming played a crucial role in the transition from human computers to electronic computers. Recognizing the potential of this emerging technology, *she taught herself and her team the programming language FORTRAN*, becoming one of the first African-American women to master this skill. Her proficiency in programming allowed her to adapt to the changing landscape of computing, ensuring that her team remained relevant and valuable to the organization.

Vaughan's contributions extended beyond her technical expertise. She actively participated in the Civil Rights Movement, advocating for equal rights and opportunities for African-Americans. Her involvement in various organizations, such as the National Association for the Advancement of Colored People (NAACP), demonstrated her commitment to social justice and equality. Dorothy Vaughan's remarkable contributions to the field of mathematics and her dedication to promoting diversity and equality have left an indelible mark on the scientific community. Her leadership, technical skills, and advocacy for equal opportunities have paved the way for future generations of women and minorities in the field of science and technology. Vaughan's legacy serves as a reminder of the importance of inclusivity and the immense potential that lies within untapped talent. Her story continues to inspire and motivate individuals to break barriers and strive for excellence, regardless of their background or gender.

# GLADYS WEST

Gladys West, an African-American mathematician and computer programmer, made significant contributions to the field of geodesy through her work at the Naval Surface Warfare Center. Despite facing numerous challenges and discrimination, West's groundbreaking research and dedication have left an indelible mark on the scientific community. We will analyze the life and achievements of Gladys West, highlighting her invaluable contributions to geodesy.

Gladys Mae Brown was born on October 27, 1930, in Sutherland, Virginia. Growing up in a segregated society, she faced numerous obstacles in pursuing her education. However, her determination and passion for mathematics led her to earn a full scholarship to Virginia State College, where she graduated with a Bachelor of Science degree in Mathematics in 1952. After completing her studies, West began her career as a teacher in Sussex County, Virginia. However, her exceptional mathematical skills soon caught the attention of the U.S. Naval Weapons Laboratory, where she was hired as a mathematician in 1956. This marked the beginning of her groundbreaking work in geodesy.

West's most notable contribution to geodesy was her involvement in the development of the *Geodetic Reference System (GRS)*. She played a crucial role in the mathematical modeling and calculations required to accurately determine the shape of the Earth. Her work involved analyzing satellite data and developing algorithms to account for various factors that affect geodetic measurements, such as gravitational anomalies and atmospheric conditions. The culmination of West's efforts was the creation of the GRS, which provided a more precise and comprehensive model of the Earth's shape. This system became the foundation for the *Global Positioning System (GPS)* that we rely on today for navigation, mapping, and various scientific applications. West's work significantly improved the accuracy and reliability of GPS technology, revolutionizing the way we navigate and understand our planet.

Despite her groundbreaking contributions, Gladys West's achievements remained largely unrecognized for many years. As an African-American woman working in a predominantly white and male-dominated field, she faced discrimination and marginalization. Her work was often overshadowed and attributed to her male colleagues, further perpetuating the erasure of her contributions. However, in recent years, West's exceptional achievements have garnered recognition. In 2018, she was inducted into the United States Air Force Hall of Fame for her contributions to GPS technology. This validation stands as a testament to her resilience and the continuing effect of her work.

Gladys West's pioneering contributions to geodesy have revolutionized our understanding of the Earth's shape and paved the way for the development of GPS technology. Her dedication, mathematical expertise, and determination in the face of adversity have left an indelible mark on the scientific community. By analyzing satellite data and developing algorithms, West's work has significantly improved the accuracy and reliability of GPS systems, benefiting countless individuals and industries worldwide. It is essential to recognize and celebrate the achievements of trailblazers like Gladys West, as they inspire future generations and challenge societal barriers.

Of course, these are only a few of the numerous scientists who have made noteworthy contributions to the fields of astronomy, physics, cosmology, astrophysics, and so on. My wish is that after hearing the personal stories of these scientists, you now have a burning desire to discover more about the countless other scientists and their contributions to the aforementioned fields. Perhaps, you will be the next great contributor to these fields, or at the least, you'll look up to the night sky and say what I said as a toddler: "I wonder what else is up there that I cannot see?"

# ABOUT THE

# AUTHOR

Prof. Robert Stewart is a retired clandestine operative who was recruited, while a student at U.C. Berkeley, into a special program for humans with Paranormal abilities. He holds a D.Sc. in Astronomy from Berkeley, and an Honorary Doctorate (Ph.D.) in World Religions from Provident University in Delaware. His fields of expertise are Martial Arts, child extraction from cults, world religions, science, and the Occult. Musician was his deep cover or camouflage life.

Prof. Robert Stewart is a multi-instrumentalist (saxophones, piano, flute, drum, vocals, etc.), composer, and producer. His two major label albums ("The Force" and "In the Gutta") were for Quincy Jones and Qwest/Warner Bros. records. He is known for his unique – personal sound and remarkably inventive improvisations declares Los Angeles Times journalist Bill Kohlhaase, as the lead tenor saxophonist on the Pulitzer Prize winning "Blood on the Fields" by trumpeter Wynton Marsalis, and as the protégé of saxophonist Pharoah Sanders. Jazz critic Jason Ankeny declared Stewart to be one of the most impressive jazz saxophonists to emerge at the end of the 20th century. Drummer Billy Higgins refers to Stewart as "perhaps the most important young artist to come along in decades."

## OTHER BOOKS WRITTEN BY PROF. STEWART INCLUDE:

The Real Mind of God: A Comparative Scriptural Analysis

Jesus: The Evidence

Gender: Issues & Solutions

Islam & Jihad (Holy War) Explained

Science and Cosmic Messengers

50 Music Compositions of Saxophonist Robert

Stewart

Buddhism For Beginners

# BIBLIOGRAPHY

Al-Khalili, Jim (4 January 2009). "The 'first true scientist'". BBC News. Archived from the original on 26 April 2015. Retrieved 2 Feb. 2024.

Bacon, Francis, *Novum Organum Scientiarum*, Nabu Press, July 22, 2011.

Bohr, Niels (1922). The Theory of Spectra and Atomic Constitution; three essays. Cambridge: Cambridge University Press.

Clarke, Arthur C. (October 1945). "Extra-Terrestrial Relays – Can Rocket Stations Give World-wide Radio Coverage?". Wireless World. Vol. 51, no. 10.

"*The Columbia Encyclopedia,*" 6th ed. New York: Columbia University Press, 2001–04.

DeGrasse Tyson, Neil (May 1, 2004). The Sky is Not the Limit. Prometheus Books.

Egginton, William, The Rigor of Angels: Max Planck unleashed a revolution in physics,

Pantheon, Delancy Place, 2023.

Farmer, Vernon L.; Shepherd-Wynn, Evelyn (2012). Voices of Historical and Contemporary Black American Pioneers. New York City.

"Gladys West | Biography, Accomplishments, Hidden Figure, GPS, Mathematician, & Facts | Britannica." February 1, 2024

Goetz, Philip W. (2007). "The New Encyclopædia Britannica". Encyclopaedia Britannica Incorporated (15th edition, Propædia ed.). Chicago, Illinois:

Gribbin, John (2013). Erwin Schrodinger and the Quantum Revolution. Trade Paper Press.

Harman, Peter M. (1998). The Natural Philosophy of James Clerk Maxwell. Cambridge University Press.

Hawking, Stephen (1994). Black Holes and Baby Universes and Other Essays. Random House.

Knox, Kevin C., Richard Noakes (eds.), *From Newton to Hawking: A History of Cambridge University's Lucasian Professors of Mathematics*, Cambridge University Press, 2003.

Lemaître, G., *The Primeval Atom - an Essay on Cosmogony*, D. Van Nostrand Co, 1950.

Liscia, Daniel A. Di. "Johannes Kepler". In Zalta, Edward N. (ed.). Stanford Encyclopedia of Philosophy.

McGraw-Hill Encyclopedia of Science & Technology (11th ed.). New York: McGraw-Hill. 2012.

*"Merriam-Webster Online Dictionary"* copyright © 2005 by Merriam-Webster, Incorporated.

Oluseyi, Hakeem; Horwitz, Joshua (15 June 2021). A Quantum Life: My Unlikely Journey from the Street to the Stars. Random House Publishing Group.

Peierls, Rudolf (1960). "Wolfgang Ernst Pauli 1900–1958". Biographical Memoirs of Fellows of the Royal Society. Royal Society.

"Rene Descartes | Encyclopedia.com" www.encyclopedia.com. Retrieved 1 February 2024.

Rosen, Edward, "Copernicus, Nicolaus", Encyclopedia Americana, International Edition, volume 7, Danbury, Connecticut, Grolier Incorporated, 1986.

Sagan, Carl; Head, Tom (2006). Conversations with Carl Sagan (illustrated ed.). Univ. Press of Mississippi.

Schutz, Bernard F. (2003). Gravity from the ground up. Cambridge University Press

Shetterly, Margot Lee (September 6, 2016). Hidden Figures: The American Dream and the Untold Story of the Black Women Mathematicians Who

Helped Win the Space Race. William Morrow.

Simmons, John (1997). The Scientific 100: A Ranking of the Most Influential Scientists, Past and Present. Secaucus, New Jersey: Carol Publishing Group.

Slayter, Elizabeth M.; Slayter, Henry S. (1992). Light and Electron Microscopy. Cambridge University Press.

Whitehouse, D. (2009). Renaissance Genius: Galileo Galilei & His Legacy to Modern Science. Sterling Publishing.

Whittaker, E. (1 November 1955). "Albert Einstein. 1879–1955". Biographical Memoirs of Fellows of the Royal Society.

Wilber, Ken (10 April 2001). Quantum Questions: Mystical Writings of the World's Great Physicists. Shambhala Publications.

Made in the USA
Middletown, DE
22 May 2025

75887914R00136